PROCESS DESIGN CASE STUDIES

Ron Scott with Norman Macleod and members of staff of the Chemical Engineering Dept, University of Edinburgh

INSTITUTION OF CHEMICAL ENGINEERS

Distributed exclusively in the USA and Canada
by VCH Publis

The information in this handbook is given in good faith and belief in its accuracy, but does not imply the acceptance of any legal liability or responsibility whatsoever, by the Institution, or by the authors for the consequences of its use or misuse in any particular circumstance.

Published by
Institution of Chemical Engineers,
Davis Building,
165-171 Railway Terrace,
Rugby, Warwickshire CV21 3HQ, UK.

Distributed exclusively in the USA and Canada by
VCH Publishers, Inc,
220 East 23rd Street,
Suite 909,
New York,
NY 10016-4606
USA.

ISBN 0 85295 276 7

Printed in England by BPCC Wheatons Ltd, Exeter.

PREFACE

This book is written for students preparing for the mandatory Institution of Chemical Engineers design project and for others with similar interests. Such students are well versed in chemical engineering theory. They are not always familiar with design procedures and practice, nor with integrating information gained from disparate sources and presenting the results to specialists of other disciplines. For the last ten years, Edinburgh University students have participated in a course of Case Study Design Exercises intended to promote these abilities. These are now presented in this book for wider use.

The author assumes that theoretical principles will be learnt from standard texts; the case studies are chosen primarily to illustrate their application to process design problems. The aim is to show how a process is designed from first principles, rather than by emulating published flowsheets. In this way, creative aspects of design may be stimulated and specific chemical engineering topics given increased relevance. At the least, the case studies may arouse interest in chemical process plants as systems designed in response to the needs of the market place, rather than as collections of disparate operations.

Studies are developed using questions and exercises whenever appropriate so that the text may be used in various ways, eg:

- by chemical engineering undergraduates before and while completing their design project;
- as a self-study course in conjunction with standard texts, students attempting questions and exercises;
- as a basis for a formal taught course along workshop or interactive lines, either in design studies or, in a more general way, to illustrate how engineers apply specific concepts;
- as a refresher for junior industrial staff involved in the development of innovative processes;
- as background reading for research and plant operating staff seeking to promote their skills and effectiveness in collaborating with process technologists and designers;

- for development into a programmed learning text with an interactive VDU responding to students' attempted answers.

USING THIS BOOK

The ultimate function of the process designer is to provide effective answers to sets of interrelated questions about process feasibility. The questions which arise in the detailed working out of any engineering design proposal are of many sorts and orders of difficulty. Some make profound demands on the designer's academic knowledge and understanding, in ways that can be met only by study (or renewed study) of scientific principles at a fundamental level; others require the careful application of pure logic, often aided by kinds of insight and lateral thinking in which designers become practised through experience.

This book is intended to contribute to the education of designers and accordingly contains much condensed experience in the form of solved case studies. The text also poses directly problems of the kind that confront designers in the course of their work, so that the reader may be encouraged to acquire the habit of finding or devising answers independently. The questions inserted as exercises have therefore an important function here; proper attention should be given to them, as well as to the narrative.

The questions in this text are of three kinds. Firstly, there are questions, designated Q., about points of fact, fundamental principle or procedure in, say, thermodynamics, which must be answered before the discussion in which they arise can continue. Such questions are, in effect, answered in the ensuing text in the specific context of the design problem in which they appear, but the reader should first pause to consider the general answer, if necessary consulting the appropriate specialist text-book to find it.

Secondly, paragraphs in these case histories sometimes end with a question mark, where some design choice or difficulty has to be resolved. Again, in the interests of narrative continuity an answer is generally supplied in the next paragraph; nevertheless, any such question is intended not to be rhetorical, but to be reflected on and answered by the reader, who should pause at the question mark to consider it — meanwhile, perhaps, covering what follows with a sheet of paper.

Thirdly, questions in the form of supplementary exercises are placed at the end of each chapter. These are intended to enlarge the reader's understanding of the foregoing material, in the way usual with text books. Outline answers are given in Appendix II.

Finally, it must be pointed out that questions in these second and third categories are usually about design choices or system arrangements. Unlike the factual or scientific questions in the first category, they may not have unique 'right' answers; on occasion, the reader may well be able to devise equally valid, or indeed 'better', answers than those suggested (though the writer does not expect this to happen very often!). It is an important characteristic of design questions that generally they do not lead to demonstrably and absolutely 'correct' answers of the kind sought for scientific questions. This and other distinctions between the designer's and scientist's activities will become clearer in what follows.

NOTES

(1) The Library and Information Service of the Institution of Chemical Engineers in Rugby, UK, offers a worldwide service for the supply of the references listed in this book.

(2) Solutions to exercises appear in Appendix II, pages 106–118.

CONTENTS

1. WHAT IS PROCESS DESIGN?

A definition of design — the creative application of fundamental principles to the fulfillment of a market need. Economic constraints on design. The need for an iterative strategy to find design(s) that best accommodate these. Setting up a database for design work. Safety considerations and the designer.

The aim of process design is to specify the most economical and effective practical procedures and plant required to produce a new product, to manufacture an existing product by new means, or, more generally, to bring about some designated material transformation on a commercial scale, so as to satisfy a market need. Design information, the body of fact with which the designer has to work, originates from research, from existing operations, or from commercial sources. This information is used in trial calculations on alternative process schemes in a search for an optimal process with minimum resource costs.

Process design is normally followed by project appraisal at management level. The first function of a designer is usually to establish to the satisfaction of the management of his company that there is an economic case for proceeding with some particular process scheme.

DESIGN OBJECTIVES AND MARKET NEEDS

What do we mean by 'design' or the act of 'designing' in the present context? We shall use the definition, 'design is the creative application of engineering theory and experience to achieve a practical objective that satisfies a market need'.

The practical objective may be a process or plant to produce a marketable product; less obviously, it may be the product itself. Examples of 'designed' products in the process industries include plastics, laminates, alloy steels and novel food products.

A product or process which does not satisfy (or generate) a market need will not normally be proceeded with, or will be short-lived.

What markets? Markets take a variety of forms. Chemical products are often intermediates between natural resources and products for use by the ultimate consumer: their markets are often within the chemical industries themselves.

In general, the closer the market is to the final consumer, the more complex the product and its quality criteria, and the higher the added value and potential reward. Thus novel products with attractive end-use properties may command a very high price: hence the current industrial interest in 'speciality' or 'performance' chemicals.

Making existing products by new processes may also open up new markets by virtue of reductions in price or improvements in quality. As an example, terephthalic acid, the intermediate for polyester manufacture, was originally produced by the oxidation of *p*-xylene by nitric acid. This process is now obsolete. A purer, and ultimately cheaper, product is now produced, for example, by oxidation of xylene by air in an acetic acid solvent with a cobalt bromide catalyst. The market for 'Terylene' expanded spectacularly when such new and cheaper processes for terephthalic acid appeared.

A new process may also serve an established market for an existing product by allowing this to be obtained from commercially attractive raw materials. In New Zealand, for instance, gasoline is being made from methanol derived from natural gas, using a novel process based on a zeolite catalyst.

A further kind of market demand is for process and plant improvements, to increase the capacity of existing installations or to improve their economics. Environmental pressures also create new demands for processes; for example to remove sulphur dioxide from power station stack gases.

These differing objectives all require successful designs for their achievement. The first aim of good design, in this context, is to satisfy market needs at least overall cost. Cost in this sense measures the total demand on available resources.

THE DESIGN PROCESS — PROBLEM SOLVING WITH CONSTRAINTS

'Available resources' include capital for plant construction, feedstocks, energy supplies, water-, sewerage-, transport-, etc services to the plant site, labour for operation and maintenance, social and environmental amenities (eg absence of hazard and pollution). If a process is to be successful it must not make unacceptable demands on such resources.

Though some of these are difficult to evaluate in monetary terms, the worth that society attaches to any resource is ultimately expressed as an equivalent cost of consumption, or resource cost. For common commodities and raw materials, this will simply be the local market value (allowing for the commercial effects of any proposed new demand); for loss of amenity and increase of hazard the equivalent costs in developed societies are constantly being raised by the pressure of public opinion and are often prohibitively high in particular areas of Western Europe and the U.S.A.

Thus, different resources have widely differing costs, which themselves vary with location and time. The process designer must take account of this in devising means of satisfying the requirement for a marketable product while minimising the total resource cost.

THE WIDER FUNCTIONS OF THE DESIGNER

It is the designer's task to solve the technical problems of, say, the synthesis of a desired product, within the (socio-)economic constraints of optimal use of resources. This, fundamentally, is what distinguishes the process designer from the pure chemist or physicist. As scientists, these latter are concerned solely with the scientific principles underlying the intended process. The designer must have wider horizons.

In the first place, he must appreciate the proposed process as a whole; not merely the chemistry or physics of the stages in which the desired product is to be formed, but also the totality of all the many steps by which the crude starting materials delivered to the plant may be transformed into packaged, or otherwise transportable, product of marketable specification.

Every one of these steps has a resource cost. For several of these possible process steps, or sequences of steps, alternatives may be devised which are functionally equivalent. Usually such alternative sequences or process routes will have different resource cost implications. The designer has then to assess the costs associated with all such alternative routes; only then will he be able to choose that of lowest resource cost, thus identifying an optimal process route or design.

DESIGN AS AN ITERATIVE PROCESS

It is generally impossible to estimate accurately in advance the resource implications even of quite minor process changes. Often, several alternative process routes, or part-routes, have to be explored and costed in detail before it becomes clear which is the most economic. Moreover, at any stage of the design process such calculations may reveal that there are unacceptably high costs associated with an entire subset of process alternatives, traceable to some earlier 'wrong turning' in the chosen process route. The systematic exploration of alternative designs has then to be restarted at an earlier point and must follow a different course.

It is thus characteristic of process design work that, after the formation of the initial scheme, it proceeds with the tentative exploration of the detailed implications of alternative technically feasible process routes. This procedure is essentially iterative; often, a path first chosen is found to lead to an unsatisfactory or questionable result, so a return has to be made to an earlier point from which another route can be explored. Many examples illustrating this iterative character of design procedures are to be found in this book; and it is worth pointing out that iteration is a feature of virtually all kinds of engineering systems design. Such design is characteristically goal-oriented (as distinct from the most orthodox kind of machine-component design, which is directed by codes of practice); the earlier choices of the design scheme are justified solely in terms of their consequences for the overall or end result. Only after these consequences have been compared with those generated by other available choices can a given design scheme be judged optimal or not.

SOURCES OF DESIGN INFORMATION

After defining market and product, the designer must begin his iterative task with an initial attempt ('first shot') at selecting the best means of meeting these objectives. This will include the processing system which converts reactants to products, the plant in which these operations are to be performed and the plan for implementing the design. Where can the designer find information enabling him to do this?

Information supporting process design may originate from many sources, including existing operations, innovative research and commercial sources. Let us examine some of these.

RESEARCH

The successful application of innovative research presents the designer with his most severe, yet most rewarding challenge. Real progress in the chemical and process industries originates in the laboratory, yet there is understandably less confidence in designs based on results from bench scale equipment than in those deriving from full scale application. How can the chemical engineer respond to this challenge?

The effective planning of research aimed at gathering design data for a new process demands an understanding of economic, as well as technical, aspects of the process. If the economic factors affecting the optimal choice of process conditions are not appreciated, laboratory data may be obtained for circumstances unrelated to those of a viable process. The chemical engineer is uniquely well placed to guide researchers away from such pitfalls. The chemical engineer should accordingly be involved in any research relevant to a new process as soon as its commercial exploitation can be foreseen. His main tasks at the research stage include devising practical and economical means of performing the operations proposed on the commercial scale, and ensuring that reliable data for scaling up and plant design are obtained for the practical circumstances that he has envisaged.

EXISTING OPERATIONS

Designs based on existing operations are likely to engender the greatest confidence. In order to develop such designs the engineer must, on the one hand, have a sound fundamental understanding of existing processes and, on the other hand, of the limits imposed by physical and chemical properties of the materials used or produced. A systematic comparison will then reveal where existing operations may be improved.

LICENCE OR PURCHASE

Present-day costs of research and process development are so high as to encourage the licensing or purchase of design information from external sources. No manufacturer designs his own nitric acid or ammonia plants today, save the very few who already have substantial experience in the field. Such plants are almost invariably purchased from an experienced contractor. In such circumstances what contribution can the purchaser's designer make?

His task is then largely to ensure that his company's standards are met, that the contractor's claims for the process are sound, and that the raw material, product and site requirements are appreciated. He may also have to make a technical and economic evaluation of competitive bids.

Buying external information about a new process while that is still at the stage of process development is a less reliable expedient, placing great demands on the chemical engineering advice available to the purchaser. To compensate for this, the financial terms should be more favourable so that the potential rewards may be greater. The buyer's chemical engineer may start by reviewing the patent literature and evaluating non-confidential assessment data provided by the vendor. If the process appears attractive, negotiation may proceed to the stage where some insight into process principles and procedure are granted under a confidentiality agreement, although this may limit the buyer's commercial or technical freedom.

STORING DESIGN INFORMATION — THE DATA BOOK

Since process design proceeds iteratively, the same basic design information will be used many times in trial calculations for different, alternative, schemes. Moreover, these trial calculations may be the work of many hands; design is usually a co-operative enterprise with contributions from different individuals. Basic data used in design must therefore be consistent throughout if trial calculations on alternatives are to yield comparable results. How can this be assured?

The basic design information is recorded together with its origin in a data book. This book is then to be the sole source of data used in all design calculations. Physical property data collected from existing operations should include details of the purity of materials involved, since the properties of commercially available intermediates and products frequently differ from those recorded in the literature for pure materials. Whenever possible, industrial designs should use properties actually measured, rather than values from the literature.

Compiling a data book is not a 'once off' exercise. It starts even before any process calculations are made, with specifications of relevant materials. It continues to grow throughout the life of the project. Some data likely to be required are indicated in Figure 1.1.

SAFETY AND HAZARDS

Safety is a resource in a special category. The cost of compromising it is nowadays prohibitive almost everywhere, and this valuation is endorsed in virtually all countries by heavy statutory penalties for unsafe practices. The safety of a processing plant is thus of paramount importance. A company's investment must be protected against legal claims for breaches of safe practice, but the social obligations of an organisation which owns and operates processing

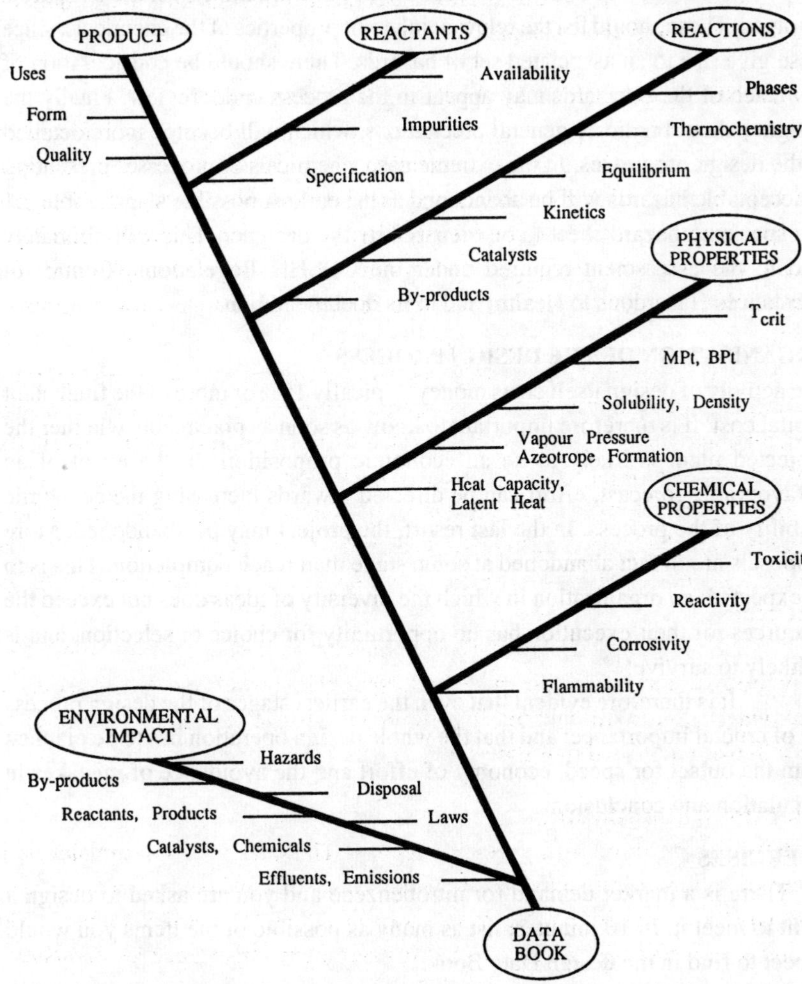

Figure 1.1 Some information for the data book.

plant go beyond the responsibilities specified by the law. Workers in the plant should not be exposed to a level of risk that could lead to accidents, ill health, etc. Of as much importance is the environment, where the product is. Users of the product are to be considered, and the local community and the environment must not suffer unacceptably. Safety aspects should therefore be considered at the earliest stage of design. How can this be done?

Data book information should include Hazard Sheets for the chemicals involved. These should list the relevant inherent properties of the chemicals, since these give rise to an associated set of hazards. There should be consideration of how each of these hazards may appear in the process under review. Finally the designer should indicate general precautions, which will become more detailed as the design progresses. In the extreme case, chemicals or processes presenting unacceptable hazards will be abandoned at the earliest possible stage. Table 1.1 is a specimen hazard sheet to be prepared by the designer. This will ultimately lead to the assessment required under the COSHH Regulations (Control of Substances Hazardous to Health) and to its documentation.

ORGANISATION OF THE DESIGN PROCESS

The activity of design itself costs money, typically 15% or more of the final plant capital cost. It is therefore important to know as soon as practicable whether the projected plant is likely to be an economic proposition. In the event of an unfavourable forecast, effort can be directed towards increasing the economic viability of the process. In the last resort, the project may be abandoned. More proposals are in fact abandoned at some stage than reach completion. This is to be expected: an organisation in which the diversity of ideas does not exceed the resources for their execution has no opportunity for choice or selection, and is unlikely to survive!

It is therefore evident that even the earliest stages of the design process are of crucial importance; and that the whole design operation has to be planned from the outset for speed, economy of effort and the avoidance of mistakes in calculation and conclusions.

EXERCISES

1.1 There is a market demand for nitrobenzene and you are asked to design a plant to meet it. In 10 minutes, list as many as possible of the items you would expect to find in the design Data Book.

1.2 You are asked to design a plant to oxidise cyclohexane with air to a cyclohexanol mixture. Prepare a Hazard Sheet for cyclohexane.

TABLE 1.1
Specimen hazard sheet: sulphuric acid

Formula:	H_2SO_4
Synonyms:	Oil of vitriol; oleum (when it contains dissolved SO_3); ROV; BOV.
Appearance:	Colourless to dark brown, viscous liquid.
Freezing point:	10.5°C. Note that observed freezing points depend on the purity and, especially, the amount of dissolved water.
Boiling point:	350°C (with decomposition).
Chemical properties:	Sulphuric acid is a strong acid, that is, it is completely dissociated in dilute solution in water. (Strength and concentration are often used synonymously, but have quite distinct technical meanings.)
Reaction with water:	Sulphuric acid is miscible with water in all proportions with considerable heat evolution; because of this, dilution of the acid by addition of water is potentially explosively violent. Dilute solutions should be prepared by cautious addition of the acid to water.
Reaction with metals:	The acid dissolves metals evolving hydrogen or sulphur dioxide. Cast iron resists strengths above about 75% w/w whilst lead is not attacked by acid of less than 80% w/w.
Considerations for plant design when sulphuric acid is used:	
Hazards arise from:	(a) Unplanned or uncontrolled mixing of acid and water (especially adding water to acid) resulting in heat generation and violent ejection from open vessels; and dilution to corrosive strengths, with damage to, or failure of, equipment; (b) Reaction with organic materials producing heat and perhaps asphyxiating sulphur dioxide; (c) Splashes and spillages causing skin irritation and chemical burns of varying severity; (d) Acid leakage onto structures, floors and drains may corrode metals and form sulphates with ceramics, leading to 'floor heave' and structural failure.

(continued overleaf)

TABLE 1.1 (continued)
Specimen hazard sheet: sulphuric acid

Emergency treatment of personal injury:	Swill affected tissue immediately with copious quantities of water, remove contaminated clothing and report for medical treatment.
Legal constraints:	*Chemical Works Regulations 1922* (SR&O 1922 No. 731) apply to manufacture or recovery of sulphuric acid.
General design and operating precautions:	
Acid is delivered to storage from road tankers at 96 – 97% strength. Vessels containing acid shall have sloped, drained acid-resisting brick floors with separate drains leading to neutralisation facilities.	
Emergency personnel drenches, protected from freezing, and eye-wash bottles shall be provided. Wearing eye protection shall be compulsory throughout the plant area. Acid-resisting protective clothing to be provided. Spillages to be copiously diluted and neutralised immediately.	

REFERENCES IN CHAPTER 1

1.1 Kirk-Othmer (Ed), *Encyclopaedia of chemical technology*, 3rd edition, Wiley, New York. For everything.

1.2 Perry, R.H. and Chilton, C.H. (Eds), 1973, *Chemical engineers' handbook*, 5th edition, McGraw-Hill, New York. Essential reference.

1.3 Sinnott, R.K., 1983, *An introduction to chemical engineering design* (Vol 6 of Coulson & Richardson's *Chemical engineering*), Pergamon, Oxford.

1.4 Weaste, R.C., Astle, W.H. and Beyer, W.H.J. (Eds), 1989, *CRC Handbook of chemistry and physics*, CRC Press Inc., Boca Raton, Fla, U.S.A. For physical properties of substances.

1.5 Baasel, W.D., 1976, *Preliminary chemical engineering plant design*, Elsevier, New York. Design of a vinyl chloride batch polymerisation plant spelled out as a case study.

1.6 Sax, N.I., 1989, *Dangerous properties of industrial materials*, 7th edition, Van Nostrand Reinhold, New York.

1.7 *Control of Substances Hazardous to Health: Control of Carcinogenic Substances, Control of Substances Hazardous to Health Regulations 1988: Approved Code of Practice*, 1988, Health and Safety Commission, HMSO, London.

1.8 Reid, R.C., Sherwood, T.K. and Prausnitz, J.M., 1987, *The properties of gases and liquids*, McGraw-Hill, New York.

1.9 *Chemical abstracts*, annual and decennial *Chemical substance* and other indices.

1.10 Rose, J.C., Wells, C.L. and Yeats, B.H., 1978, *A guide to project procedure*, IChemE, Rugby. Part 1 explains who does what, from project conception to operating plant. Note particularly page 4 explaining the work of a process engineering group, Chapter 6 on process design, and the appended vessel drawings.

1.11 Landau, R. and Cohan, A.S. (Eds), 1966, *The chemical plant from process selection to commercial operation*, Reinhold, New York. The inside story by the chief and staff of an organisation that has been among world leaders in designing plants based on innovative research.

1.12 Resnick, W., 1981, *Process analysis and design for chemical engineers*, McGraw Hill, New York.

1.13 Ray, M.S. and Johnston, D.W., 1989, *Chemical engineering design project — a case study approach*, Gordon & Breach, New York.

2. MARKET NEEDS AND DESIGN OBJECTIVES

Market requirements for product specification effectively determine the design of the purification stage of the manufacturing process. Furthermore, the required product purity may be such that the reactor outflow must itself be of high purity if feasible purification processes are to deliver a product meeting the customer's specification. Thus market requirements may also determine the minimum conversion acceptable in the reactor stage.

These principles are exemplified by the two cases of methyltertiarybutyl ether (MTBE) and cyclohexane manufacture, where difficulties of purification to specification impose constraints on reactor performance —in the second case to an extreme degree.

Process design is market-led. The market imposes volume, cost and quality objectives for a product, derived from the scale and nature of the customer's use. Product quality is affected by impurities originating in the raw material chosen and arising from the manufacturing process selected, and by the means used for contaminant suppression or removal. The market thus influences raw material selection, the choice of process and the separation procedures used, as well as the scale of the operations.

THE MARKET, THE PRODUCT AND THE PROCESS

As shown in Chapter 1, the purpose of a manufacturing process is to convert reactants to products which satisfy a market need. 'The Customer is King'; the price he is willing to pay for the product — its value to him — depends on the extent to which it meets his specification.

A successful process is one which enables the manufacturer to achieve a satisfactory product at a cost to him lower than its sale value to the customer. After recovering all operating costs there must be an adequate operating margin between these and the selling price of the product to repay research costs and capital investment.

This is the basis of an essential economic screening test for new products and processes. To apply it, the designer needs to know the prospective selling price of his product. Estimating this requires detailed knowledge of the customer's operations and the alternatives available. The price of existing products is additionally determined by competition from other makers and alternative processes. Designers seek advice in these areas from market specialists.

The designer can, however, himself make preliminary cost estimates for possible processes. If no better information is available he assumes, in the first instance, near stoichiometric yield. He then multiplies materials cost by a factor of perhaps two to allow in a rough preliminary way for other operating costs. These can later be more satisfactorily estimated once an outline process is available. Unless the proposed process leads to an adequate margin between such preliminary estimates of manufacturing cost and the market specialist's predicted selling price, the designer must proceed with caution!

PRODUCT QUALITY

Success in meeting the customer's specification depends on removing or suppressing impurities originating in raw materials or in the process itself. Raw materials should therefore be chosen and processes structured to meet these requirements. A procedure for accomplishing this is outlined in Figure 2.1 (see page 14).

Here, identification of the market is followed by an appraisal of desired product specification (purity) as a function of end use and competition from other suppliers. Consideration of possible manufacturing routes then leads to the need to assess impurities and purification options associated with each. Means of separating the desired product from those of possible side reactions are then examined. The difficulties of removing contaminants to the required extent must be weighed against the difficulties of suppressing their formation by suitable choice of reactor conditions. Hence arises the need to establish design objectives for the reaction stage. The feasibility of meeting these objectives by choice of reaction conditions or reactor design is explored in Chapter 3.

The procedure is exemplified in the first case study, which illustrates the maxim: 'design is about choice'. It concerns a product, the gasoline anti-knock blending agent MTBE, which can be made to market specification by more than one combination of reaction conditions and separation schemes.

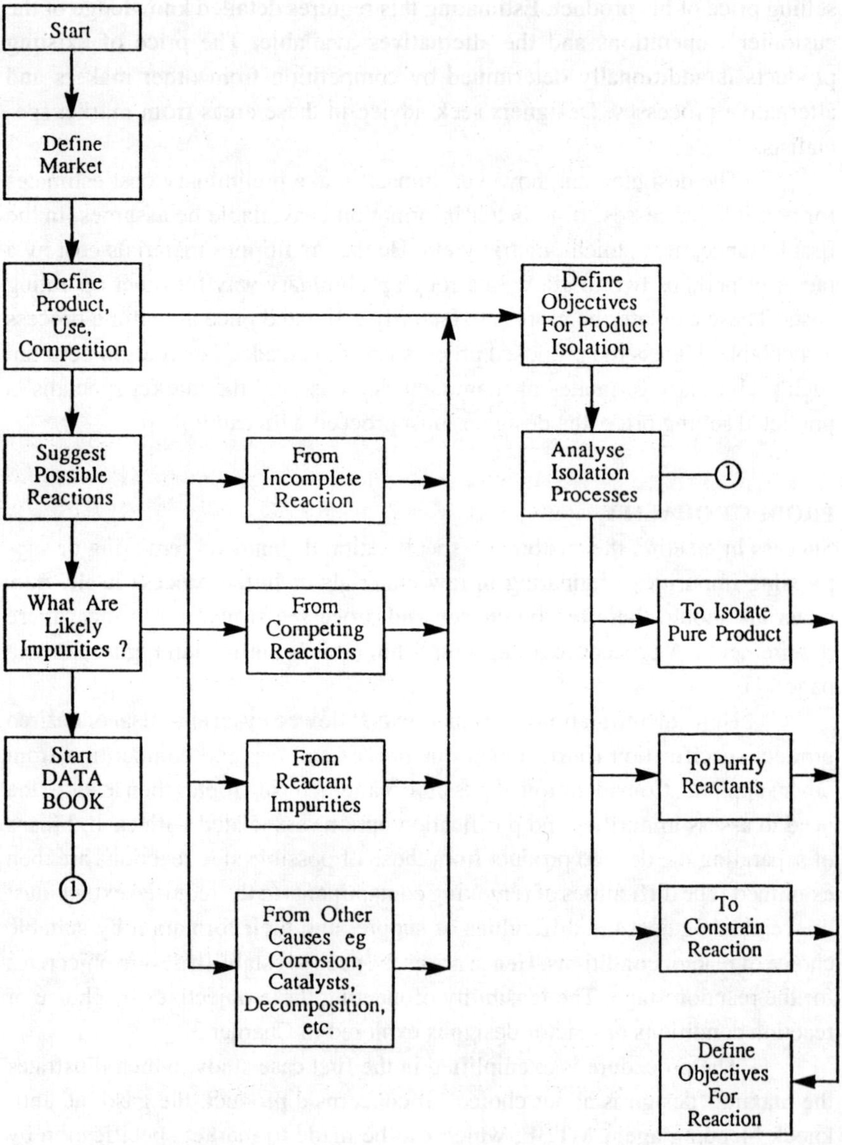

Figure 2.1 Flowchart for preliminary process analysis.

CASE STUDY 2.1 — MTBE PRODUCTION

THE MARKET

At the high temperature and pressure reached during the compression stroke in internal combustion engines, the gasoline/air mixture is liable to ignite before the spark. This is called 'knocking', from the sound produced by the pressure wave out of phase with the piston motion. It reduces power output and may damage the engine. The ability of gasoline to resist this premature ignition is measured by its Research Octane Number (RON). Unblended gasolines show a RON around 90 (regular), but this is currently raised by additives or blending agents to at least 93 (2 star petrol) or to about 95 (unleaded) or 97 (4 star), allowing the use of higher engine compression ratio and greater thermal efficiency. There is thus an important market for such agents.

THE PRODUCT, ITS USE AND THE COMPETITION

The established anti-knock additive is tetraethyl lead (TEL), but this is toxic and also poisons those exhaust catalysts designed to remove unburned hydrocarbons. Other suitable compounds are methanol and aromatics. A blend of gasoline with up to 10% methyltertiarybutyl ether (MTBE), as permitted by BS 7070 of 1988, gives a satisfactory knock rating. Since MTBE must be present in substantial concentration, it is described as a blending agent; TEL is effective at a few parts per million and is classed as an additive.

There is also a demand for MTBE for use as a reagent in the chemical industry. In Japan it is used as an intermediate in methylmethacrylate production.

POSSIBLE REACTANTS AND REACTIONS

Isobutene reacts catalytically with methanol in the reaction:

$$CH_3OH + (CH_3)_2C{:}CH_2 \rightarrow (CH_3)_3COCH_3$$

Isobutene is available as a 50% w/w component of the mixed C_4 stream originating from a steam naptha cracker after removal of butadiene. An alternative source is the C_4 stream from fluid bed cracking of refinery fractions, which contains about 16% isobutene.

POSSIBLE IMPURITIES

Impurities in the product may arise from the following sources:

(a) From incomplete reactions.

Isobutene: Miscible and more volatile, so acceptable within limits.

Methanol: Also has anti-knock properties. Acceptable within limits.

(b) From competing reactions.
Dimers, trimers and, at high temperatures, oligomers of lower volatility; these are of limited acceptability, being liable to polymerise in the engine fuel system.
(c) From impure reactants.
Other hydrocarbons in the isobutene stream are mainly C_4s together with traces of C_3s and C_5s. Water in the feed will react, yielding tertiary butanol which is miscible and has anti-knock properties; water in the product stream must be strictly limited because of its low solubility.
(d) From other sources.
Other impurities may be discovered as the design progresses, derived, for instance, from materials of construction, catalysts or other auxiliary substances necessary for the implementation of particular process schemes.

Table 2.1 gives examples of product specifications currently offered by different manufacturers, evidently acceptable to end users.

DESIGN OBJECTIVES FOR PRODUCT ISOLATION
The overall design objective is to produce product that meets the minimum commercial specification by sufficient reduction of impurity sources (a) – (d). We consider (a) and (b) in the next chapter. To counteract (c) there are two possibilities:

TABLE 2.1
Commercial specifications for MTBE

	wt%	wt%	wt%
MTBE	99.35*	98.4	98.6
Methanol	0.4	1.2	< 5 ppm
C_4	0.05	0.2	< 0.2
Di-isobutene	—	—	< 0.2
Tert-butanol	0.2	0.2	1.0
Water	< 2 ppm	< 10 ppm	< 5 ppm

* For chemical use

TABLE 2.2
Composition of mixed C_4-stream

	Boiling point, °C	Composition, wt%
iso-butane	–11.9	1
isobutene	–6.9	50
1-butene	–6.3	28
n-butane	–0.5	10
trans-2-butene	0.8	7
cis-2-butene	3.7	4

(1) Purify reactants. Isobutene originates in a mixed C stream which, when a steam/naphtha cracker is the source and valuable butadiene has been removed, has the composition shown in Table 2.2.

❐ Q. *Is it feasible to isolate isobutene as a pure feed stream before reaction? What property differences might be used to do this?*

Distillation is usually the cheapest way to separate hydrocarbons. Isobutene is a middle fraction in the C_4 stream so that at least two stages of distillation are needed to isolate it. The first, removing isobutene, the more volatile components and almost 50% of the mass, has isobutene and butene-1 as key components. These have a bubble point difference of only 0.6°C and relative volatility of 1.01. Some hundreds of plates would be required for their separation, which is therefore impracticable. The whole of the stream must therefore be passed through the reactor and we are left with possibility (2).

(2) Isolate product from impurities. The reactor outflow will contain some or all of the following components:

(i) Unreacted C_4s. For separation by distillation all these may be treated as a single component with the properties of, say, *cis*-butene-2.

(ii) MTBE.

(iii) Methanol.

(iv) Tertiary butanol, formed when isobutene reacts with any trace of water present in the feed.

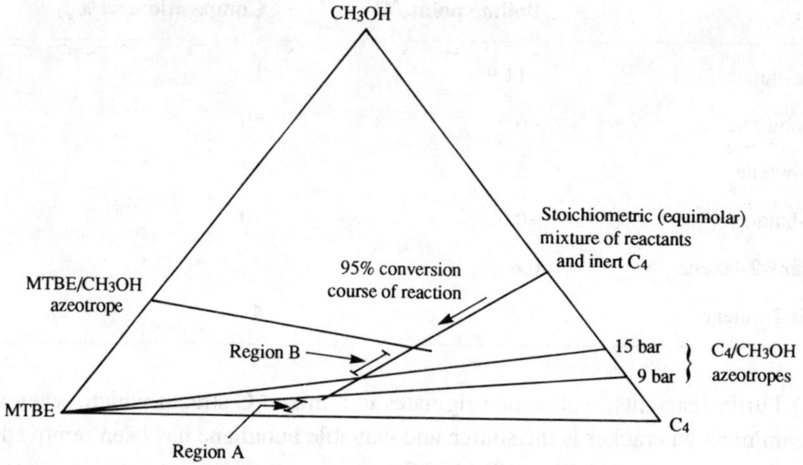

Figure 2.2 MTBE/methanol/C_4 azeotropes.

(v) Di- and tri-isobutene and other high boilers formed in the reaction.

Differences in volatility and miscibility are indicated in the literature. These are shown in Table 2.3 as they would appear in the data book. Distillation design rules usually suggest removing the most volatile components first. At least four separations are necessary to separate five components. In this case, however, distillation will not necessarily yield these components as pure fractions because of the existence of azeotropes, as indicated in the notes to Table 2.3.

The following fractions may be expected:

- C_4/methanol azeotrope.
- C_4s, unless entirely removed in the first fraction.
- MTBE/methanol azeotrope.
- MTBE.
- Methanol, unless previously removed in the azeotropes.
- High boilers, ie components (iv) and (v) above.

Figure 2.2 shows schematically (not to scale) the essential features of the azeotropic behaviour for C_4/methanol/MTBE mixtures from proprietary information disclosed in Reference 2.5. From this, two possibilities emerge; the

TABLE 2.3
Extracts from the data book

Component	Formula	Molecular weight	Boiling point, °C	Critical temperature, °C
isobutane	C_4H_{10}	58.1	−11.9	135
iso-butene	C_4H_8	56.1	−6.9	145
1-butene	C_4H_8	56.1	−6.3	147
n-butane	C_4H_{10}	58.1	−0.5	152
trans-2-butene	C_4H_8	56.1	0.8	156
cis-2-butene	C_4H_8	56.1	3.7	163
MTBE	$C_5H_{12}O$	88.1	55.2	228
methanol	CH_4O	32.0	64.6	240
tert-butanol	$C_4H_{10}O$	74.1	82.8	235
di-isobutene	C_8H_{16}	112.2		
tri-isobutene	$C_{12}H_{24}$	168.3		

Notes: (1) Boiling points and critical temperatures are given (somewhat unusually) in degrees Celsius here.
(2) The isobutene oligomers are not pure substances and therefore available physical data must be used with caution. There is a methanol/MTBE azeotrope of composition about 15:85 w/w varying with pressure. Methanol also forms an azeotrope with the C_4 hydrocarbons whose composition varies with pressure and MTBE content. Outline phase boundary composition data are given for these mixtures in Figure 2.2.
(3) Solubility data. Water dissolves 4.0% by weight of MTBE and MTBE dissolves 1.3% by weight of water. Methanol and water are miscible in all proportions. The hydrocarbons are insoluble in water.

product stream from the reactor may contain a proportion of unreacted methanol either in region A below or region B above the azeotrope composition.

SCHEME A

With a feed to the separation section whose composition falls in the region A, a distillation would yield MTBE and high boilers as the bottoms product. From the overheads it would not be possible to recover all C_4s free from methanol because the azeotrope is the most volatile fraction. Does this matter? Is it necessary to separate methanol from the residual C_4s?

Unless methanol is removed, it is questionable whether the C_4s are of any commercial value other than as fuel. Also valuable methanol is lost. What property might be used for their separation?

Table 2.3 indicates that there are differences in miscibility which might serve to separate the components. Thus methanol may be removed from C_4s by water extraction and phase separation. The C_4 stream will require drying while methanol and water can be separated by distillation, the bottoms being recycled to the extraction stage. These operations are shown in Figure 2.3 as Distillation 1 in block 400, Distillation 2 in block 500, and C_4 Drying in block 900.

The next design objective is to separate the bottoms from Distillation 1, containing methanol, MTBE and high boilers. The concentration of methanol in the bottoms stream depends on the compositions of the reactor outflow and of the C_4/methanol azeotrope. The latter depends on the column pressure. If, at a practicable value of this, all the methanol can be removed overhead in Distillation 1, only MTBE and high boilers remain and may be separated by Distillation 3 in block 800. The minimum number of three separation steps for isolation of MTBE from methanol, other C_4s and high boilers is increased to five by the presence of the azeotrope and the introduction of water to break it. Calculations which will not be reproduced here suggest that the mass balance will support Scheme A provided that the mol fraction of methanol in methanol plus C_4s leaving the reactor is not more than approximately 0.087 (ie 5% by weight). This appears to be the principle of the process operated by Phillips Petroleum using feedstock from a fluid bed catalytic cracker.

SCHEME B

If the methanol concentration exceeds this, feed to the separation block corresponds to region B on Figure 2.2. Distillation 1 then yields methanol, MTBE and high boilers as raffinate. How might this be separated?

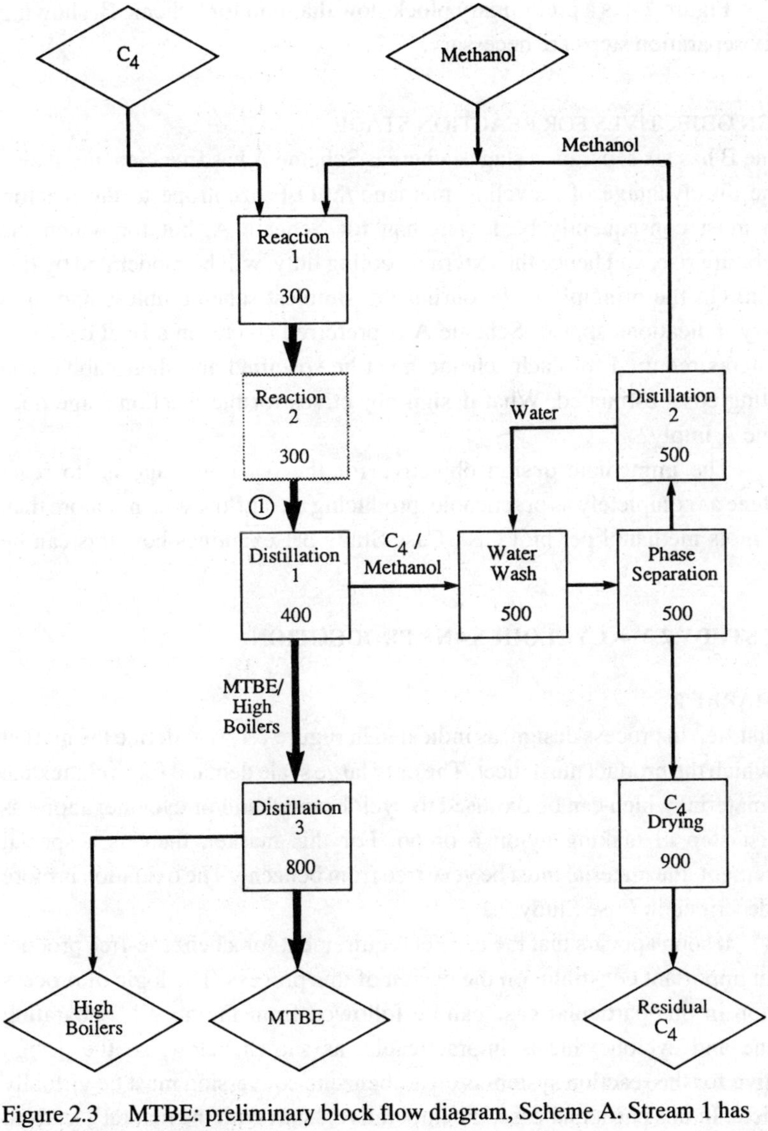

Figure 2.3 MTBE: preliminary block flow diagram, Scheme A. Stream 1 has composition A in Figure 2.2.

Two distillations are required to separate this three component mixture. An additional distillation yields methanol/MTBE azeotrope overhead for recycle to the reactor with MTBE and high boilers as raffinate for final separation.

Figure 2.4 is a preliminary block flow diagram for Scheme B, showing that six separation steps are necessary.

DESIGN OBJECTIVES FOR REACTION STAGE

Scheme B has six separation stages whereas Scheme A has five. Scheme B also has the disadvantage of recycling methanol/MTBE azeotrope to the reactor, which must consequently be larger than for Scheme A, but for which the temperature rise, and hence the external cooling duty, will be moderated by this dilution. On the principle of favouring the simplest scheme unless and until contrary indications appear, Scheme A is preferred. To reach a final decision, plant items required for each scheme must be specified and their capital and operating costs estimated. What design objectives for the reaction stage does Scheme A imply?

The immediate design objective for the reaction stage is 'to react isobutene as completely as practicable, producing an outflow with not more than 0.095 mols methanol per mol C_4s'. Case Study 3.1 examines how this can be done.

CASE STUDY 2.2 — CYCLOHEXANE PRODUCTION

THE MARKET

The first step in process design, as indicated in Figure 2.1, is to define the market need which the product must meet. The only large scale demand for cyclohexane is for material which can be oxidised to cyclohexanol and/or cyclohexanone as the first step in making nylon 6 or 66. For this market, there is a special requirement; the material must be very free from benzene. The oxidation is more fully described in Case Study 9.1.

It soon appears that the market requirement for a benzene-free product puts an important constraint on the design of this process. The logic of process selection in this particular case can be followed from Figure 2.1. Separating benzene and cyclohexane is impracticable, as shown below, so the design objective for the reaction system is clear; benzene conversion must be virtually complete. In this particular case, the imperative, 'start with the market', leads to a unique choice of reaction scheme.

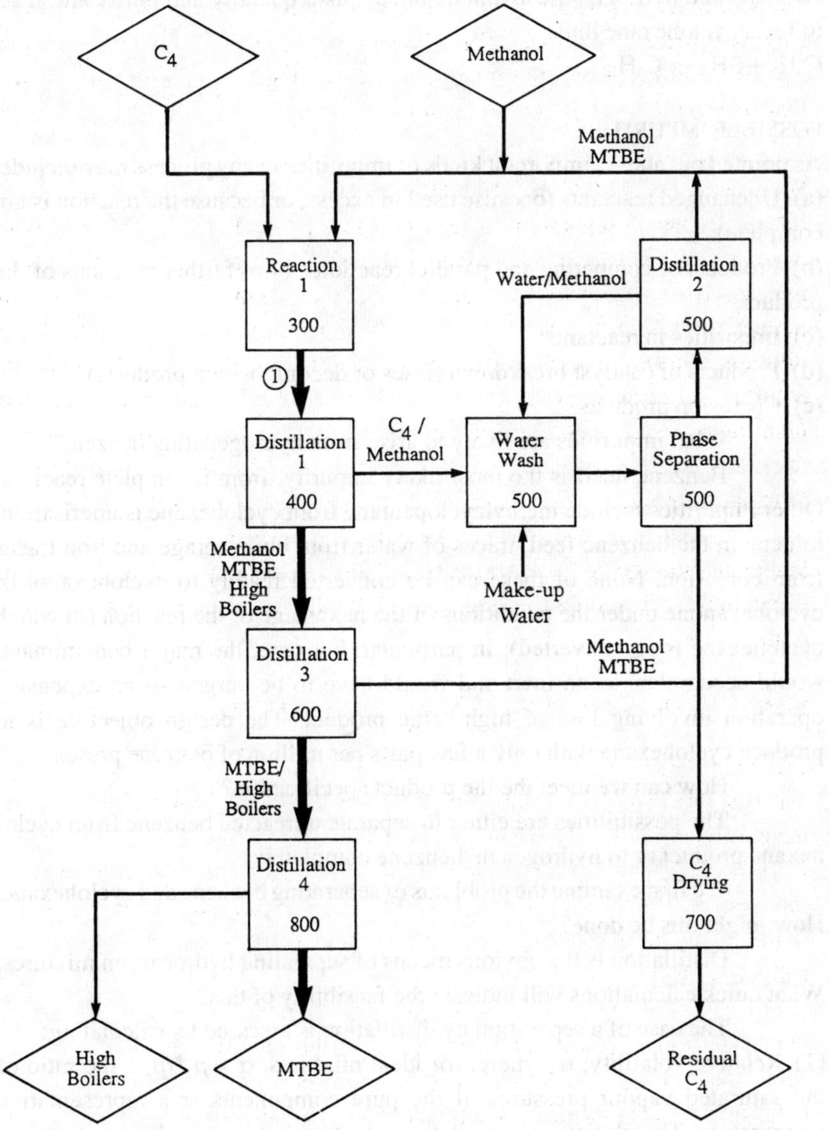

Figure 2.4 MTBE: preliminary block flow diagram, Scheme B. Stream 1 has composition B in Figure 2.2.

POSSIBLE REACTANTS AND REACTIONS

Benzene and hydrogen are available in adequate quantity and purity and react to form cyclohexane thus:

$$C_6H_6 + 3H_2 \rightarrow C_6H_{12}$$

POSSIBLE IMPURITIES

As pointed out above, important kinds of impurities in any process may include:

(a) Unchanged reactants (because used in excess, or because the reaction is not complete);

(b) Products of competing and parallel reactions, or of further reactions of the product;

(c) Impurities in reactants;

(d) Products of catalyst breakdown (fines or decomposition products);

(e) Corrosion products.

What impurities are likely to arise from hydrogenating benzene?

Benzene itself is the most likely impurity, from incomplete reaction. Other impurities include methylcyclopentane from cyclohexane isomerisation, toluene in the benzene feed, traces of water from bulk storage and iron traces from corrosion. None of these can be converted readily to cyclohexanol or cyclohexanone under the conditions of the next stage of the reaction (in which cyclohexane is so converted); in particular, benzene, the major contaminant, would accumulate as an inert and would have to be purged — an expensive operation involving loss of high-value product. The design objective is to produce cyclohexane with only a few parts per million of benzene present.

How can we meet the the product specification?

The possibilities are either to separate unreacted benzene from cyclohexane product or to hydrogenate benzene completely.

We first examine the problems of separating benzene and cyclohexane. How might this be done?

Distillation is the obvious means of separating hydrocarbon mixtures. What quick calculations will indicate the feasibility of this?

The ease of a separation by distillation is assessed by calculating:

(1) Relative volatility, α, where, for ideal mixtures, $\alpha = p_1^*/p_2^*$, the ratio of the saturated vapour pressures of the pure components at a representative temperature. The obvious course is to evaluate this for our mixture at atmospheric pressure, although values of α vary with temperature and hence with system pressure.

❒ Q. *How can we estimate the relative volatility of benzene and cyclohexane assuming that mixtures are ideal?*

Using Antoine coefficients at 80.7°C, the atmospheric boiling point of cyclohexane, the benzene vapour pressure is 770.5 mmHg, so that the relative volatility is 1.014. Some mixtures which are separated satisfactorily commercially show the following relative volatilities, varying with temperature:
Benzene/EDC, 1.11
Benzene/toluene, 2.3 – 2.6
Water/ethylene glycol, 20 – 100.
Separating benzene and cyclohexane might be difficult! A more accurate picture is given by calculating:
(2) Theoretical stages required at total reflux. The Fenske Equation gives this as *S*, where
$S \log \alpha = \log(x_1/x_2)_{tops} - \log(x_1/x_2)_{bottoms}$

Cyclohexane, unusually for a saleable product, is taken as bottoms. Its benzene concentration is to be minimal, say 0.1% for a trial design. At this concentration, a plant consuming 100 kt/year cyclohexane will accumulate 100 t/year benzene, which seems reasonable.

What is the minimum number of theoretical plates, N (=S –1), required to separate benzene and cyclohexane at total reflux?

We may assume benzene concentration in the product to be 0.001. Benzene recovered overhead is recycled for hydrogenation so that cyclohexane content does not appear critical; $(x_1)_{tops}$ may be assumed to be 0.5. If the number of actual plates is some 4× theory and plate spacing is approximately 0.5 m, is the distillation practicable?

Use of these values shows N to be 497 with a suggested column height of c.1000 m. This is clearly impracticable! Assumptions made to accelerate the progress of preliminary design need to be confirmed later. More accurate VLE data are reported in Reference 2.1 but relative volatility is still low on that basis at 1.016, so that distillation is still impracticable. Reference 2.2 reports an azeotrope but the Fenske Equation indicates that a minimum of 80 stages are required for its separation. Distillation is therefore to be avoided.

DESIGN OBJECTIVE FOR THE REACTION STAGE

The design objective for the reaction stage is therefore to produce an outflow which already meets product specification for benzene content, ie conversion is to be not less than 99.9%. How this can be done is discussed in Case Study 3.2.

EXERCISES

2.1 MTBE Scheme B allows recirculation of methanol/MTBE azeotrope to the reaction stage. More methanol in the reactor outflow then means less MTBE isolated as product. Devise an alternative separation procedure for this case, different from that suggested here, and draw a preliminary block flow diagram for it.

2.2 The C_4 stream in Case Study 2.1 originates from the primary distillation column of a steam naphtha cracker. What impurities other than C_4s are likely to be present? Where will these impurities emerge from the proposed separation process? Will they affect performance significantly?

REFERENCES IN CHAPTER 2

2.1 Scatchard, G., Wood, S.E. and Mochel, J.M., 1939, *J Phys Chem* 43: 119–130.

2.2 Weaste, R.C., Astle, M.J. and Beyer, W.H. (Eds), 1986, *CRC Handbook of chemistry and physics*, CRC Press Inc., Boca Raton, Fla. 67th edn., p. D19.

2.3 Pecci, G. and Floris, T., Dec. 1977, *Hydrocarbon Processing,* 98–102.

2.4 Obenaus, F. and Droste, W., *Erdol u. Kohle* 33 (6 June 1980) 271–275, and 40 (15 June 1987).

2.5 Ancillotti, F. *et al.*, 1981, *UK Patent 1,587,457.*

2.6 *Specification for unleaded petrol (gasoline) for motor vehicles*, BS 7070, 1988, amended June 1990, British Standards Institution.

3. CHEMICAL EQUILIBRIUM AND REACTOR CHOICE

The next stage in constructing a process design scheme involving chemical reaction is to determine under what conditions the reactor can function so as to satisfy the outflow requirements now arrived at. In the first place this is a thermodynamic matter, as appears from further consideration of the MTBE and cyclohexane cases.

In Chapter 2 we considered the specifications of two products, MTBE and cyclohexane, and examined for each the possibility of reducing undesirable impurities to required levels in the separation and purification stages; we thus set design objectives for the reaction stages. We must now establish whether it is feasible to design reactors capable of attaining these objectives.

SOME GENERAL PRINCIPLES OF REACTOR SELECTION

At an early stage in the design of any process involving chemical reaction, possible constraints must be identified and then explored to establish operating conditions for the reactors. Considerations of reaction equilibrium (the thermodynamics) come first, before we ask about the reaction rate (the kinetics). The logic of this approach is set out in Figure 3.1 (see page 28).

In following this scheme it can be seen that important questions to be answered are: whether we are to use gas phase or liquid phase reactions, whether the reaction is endo- or exothermic, and whether or not the reaction is reversible.

CHOICE OF REACTANT PHASE

In general, many advantages are offered by liquid phase, rather than gas phase, operation. For the required output the reactor will be smaller, not only because liquids are more dense, but also because they have greater heat capacities and thermal conductivities than gases — factors enhancing heat transfer. Small equipment size is an advantage in relation to what is known as process intensification; both power requirements and capital costs will be lower.

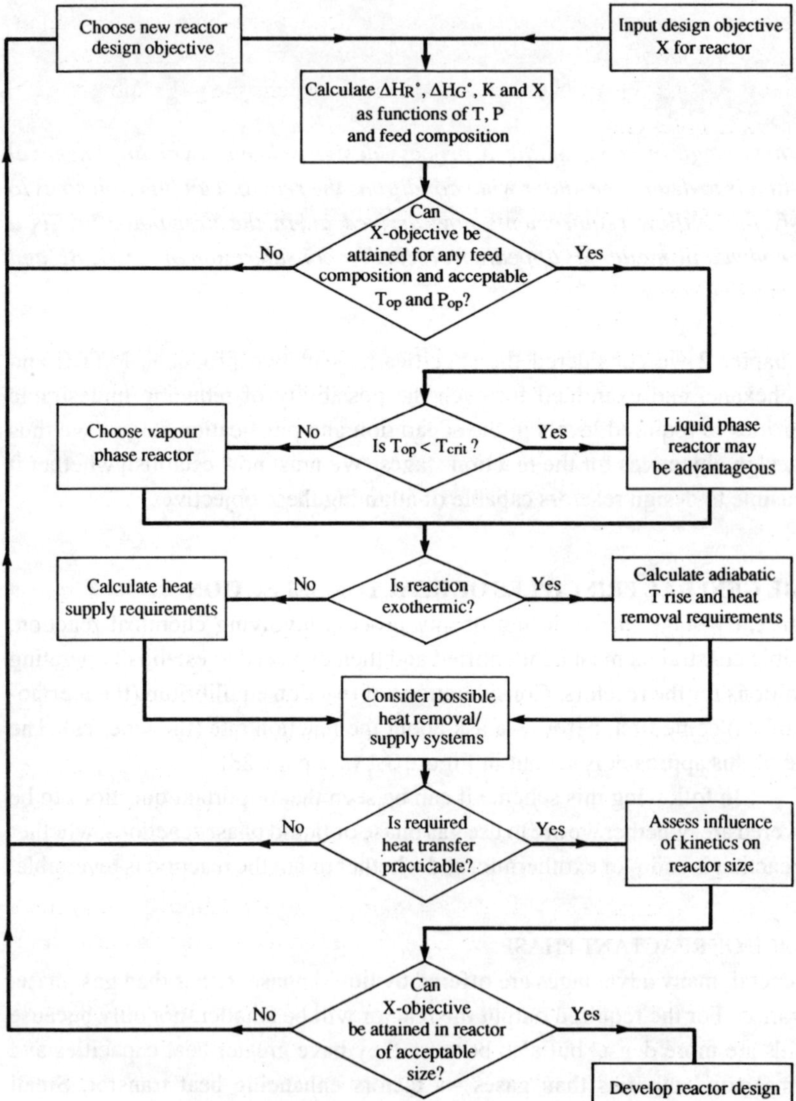

Figure 3.1 Flowchart for reactor design.

There may, however, be significant disadvantages, such as enhanced corrosion and catalyst losses. More fundamentally, in a liquid phase system all operating conditions must fall within the two phase region; if the critical temperature is not much above the desired reaction temperature, high operating pressures — potentially hazardous and expensive to contain — may be necessary, particularly if one of the reactants is a non-condensible gas required to be at high partial pressure.

QUESTIONS OF THERMOCHEMISTRY AND REVERSIBILITY

Two fundamental thermodynamic questions are to be asked:

- What is the heat of reaction?
- Is the reaction reversible?

For the particular cases of MTBE and cyclohexane production, calculations show that the reactions under consideration are both exothermic and reversible for both these products:

$$C_4H_8 + CH_3OH = (CH_3)_3COCH_3$$

$$\Delta H_R = -37.6 \text{ kJ/mol (gas phase)}$$

$$C_6H_6 + 3H_2 = C_6H_{12}$$

$$\Delta H_R = -205 \text{ kJ/mol}$$

The heat of reaction might be utilized in both cases to generate a substantial temperature rise, enhancing the reaction rate, but in both cases our situation is complicated by the reversibility. We encounter a well known optimisation problem; at high temperatures the reaction rates are very high but equilibrium conversion is low, whilst at low temperatures the rates are low although the equilibrium conversion can be high. In a batch reactor, on admixture of the pure feeds the reaction accelerates due to the exothermic temperature rise, and then slows down until the net rate is zero at an equilibrium conversion which is lower (ie further to the left) the higher the system temperature. Thus the next stage in our process analysis procedure is the calculation of equilibrium constants and a comparison of the equilibrium conversion with our design objectives.

It is worth remembering that, as Figure 3.1 shows, entirely different considerations would apply if these exothermic reactions were irreversible. We would then be primarily concerned with the influence of reactor design on conversion and with heat transfer and control.

CASE STUDY 3.1 — MTBE PRODUCTION

Consideration of the reaction of isobutene with methanol illustrates the use of these general procedures. Case Study 2.1 defined the design objective for the reaction stage: to react isobutene as completely as practicable, producing an outflow with no more methanol than can be removed as the azeotrope with C_4s. It is shown in Chapter 2 that MTBE can still be isolated if the methanol concentration is greater, but more operations are necessary.

One soon encounters the problem that some of the required thermodynamic or kinetic data are not available. Until this has been remedied in the laboratory, estimates must be used, such as thermodynamic data relating to closely similar compounds. A further complication is that the 'isobutene' is in fact a multi-component stream of C_4 hydrocarbons. (See Table 2.2, page 17.)

The feasibility of maintaining the reactants in the liquid phase can be tested by calculating pseudo-critical constants by established methods such as those using weighted van der Waals constants. In the present case, the best that we can do is to use simple weighted averages relating to approximate compositions. In the absence of further information, the C_4 stream might be approximated by an equimolar mixture of isobutene and *n*-butane. The critical constant values then suggest that we take a value of 41 bar for the pseudo-critical pressure. System pressures, including any contribution from inerts, can be checked using Raoult's Law. Vapour pressures for pure substances can be found from tabulated Antoine coefficients, but extrapolation outside the range of validity is likely to be necessary and must be done with caution, especially near the critical point.

In Appendix I we have calculated the equilibrium constant for the formation of MTBE from an equimolar mixture of isobutene and methanol by estimation of the standard Gibbs free energy change of reaction at various temperatures. Using material balance equations for the components, the equilibrium conversion can be plotted as a function of temperature, with or without inerts in the form of *n*-butane. It is also shown in Appendix I that, in the absence of some means of heat removal, there would be a substantial, and indeed unacceptable, temperature rise in the course of the reaction. The results of these calculations are given in Figure 3.2, which shows the equilibrium conversion of the isobutene vs temperature for an originally equimolar mixture of isobutene, isobutane and methanol for the limiting case of no dilution by further inerts or recycle.

For such a system we see that even if suitable catalytic reaction rates could be achieved at temperatures as low as 300°K or 30°C, because of the

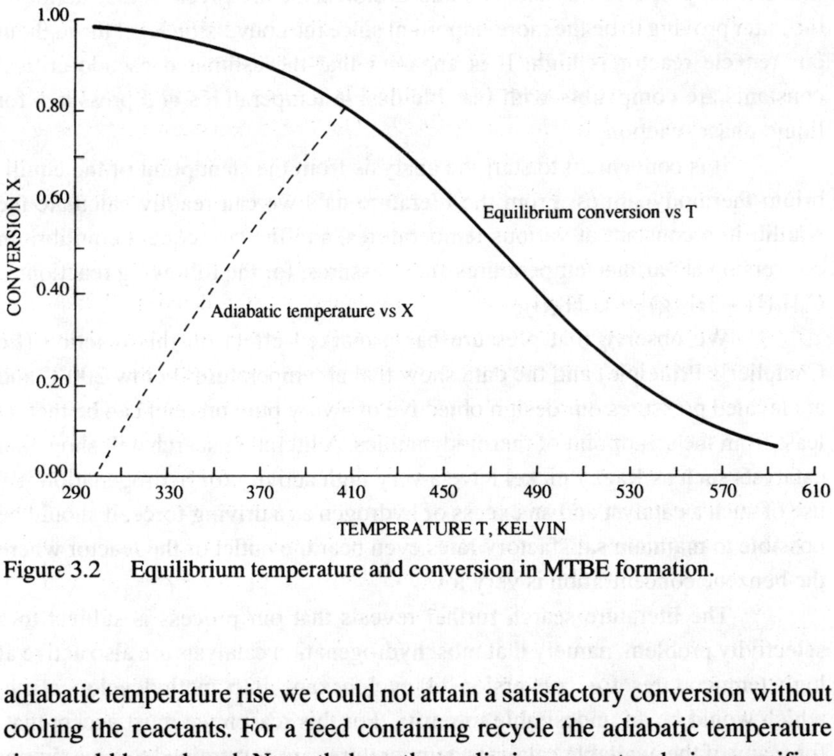

Figure 3.2 Equilibrium temperature and conversion in MTBE formation.

adiabatic temperature rise we could not attain a satisfactory conversion without cooling the reactants. For a feed containing recycle the adiabatic temperature rise would be reduced. Nevertheless, the preliminary reactor design specification must call for a reactor with internal cooling; alternatively, if an adiabatic reactor is used it must be staged, with intercoolers. For a feed approximately equimolar in isobutene, butane and methanol, stoichiometric considerations discussed in Appendix I show that, if the methanol/C_4 mol ratio in the reactor outflow is to be kept below 0.095, as recommended in Chapter 2, the conversion should be not less than 95%. The maximum allowed temperature in the product stream, consistent with a conversion of 95%, is 343°K, or 70°C, but we should note that higher temperatures can be allowed upstream in the reactor to gain an enhanced overall reaction rate. A literature search will show that certain sulphonated polystyrene resins are suitable catalysts for the reaction.

CASE STUDY 3.2 — CYCLOHEXANE PRODUCTION

PRELIMINARY ANALYSIS

The analysis for this case follows the procedures used in the MTBE Case Study. The critical properties of benzene and cyclohexane are given in the literature, the latter proving to be the more important since the conversion level throughout our recycle reactor is high. It is apparent that the estimated pseudo-critical constants are compatible with feasible design temperatures and pressures for liquid phase reaction.

It is convenient to start the analysis from the standpoint of the equilibrium thermodynamics. From the literature data we can readily calculate the equilibrium constant at various temperatures, and the consequent equilibrium conversion at various temperatures and pressures, for the following reaction:

$C_6H_6(l) + 3H_2(g) \rightarrow C_6H_{12}(l)$

We observe that pressure has a marked effect on this reaction (Le Chatelier's Principle) and the data show that at temperatures below 200°C and at elevated pressures our design objective of a very pure product can be met, at least from the standpoint of thermodynamics. A literature search will show that catalysts such as Raney nickel have a very high activity for hydrogenation. By use of such a catalyst and an excess of hydrogen as a driving force, it should be possible to maintain satisfactory rates even near the outlet of the reactor where the benzene concentration is very low.

The literature search further reveals that our process is subject to a selectivity problem, namely that most hydrogenation catalysts are also active at high temperatures for conversion of cyclohexane into methylcyclopentane, which would be an undesirable impurity. For this reason we must accept that, for many of the available catalysts, temperatures are constrained to a maximum of 200°C throughout the reactor. The design treatment is, in these circumstances, little different from that of an irreversible exothermic reaction with a prescribed temperature limit. For a reaction subject neither to equilibrium (reversibility) nor temperature limits as such, we might have preferred yet higher temperatures from considerations of the economics of heat recovery.

The very high heat of reaction suggests that we must pay special attention to heat removal in the design. For example, full use must be made of evaporative cooling in the product stream, cold recycle and perhaps cold-shot cooling, ie the injection of part of the (cold) feed into the reactor at one or more points downstream of the main feed entry. Since we require high conversion, such procedures, which imply backmixing, can only be considered because of the availability in this particular case of high activity catalysts.

BLOCK FLOW DIAGRAMS

The preliminary investigation of the thermodynamics of the hydrogenation of benzene suggests that it may be possible to achieve the desired specification for the product by complete reaction using an excess of hydrogen and a reaction temperature not exceeding 200°C. At this point a block flow diagram of process stages can be produced for later development.

When constructing block flow diagrams it is useful to ask four questions about each stage:

(a) What is the objective?
(b) What are the constraints within which this must be achieved?
(c) What is the consequent range of possible operating conditions?
(d) Are there likely to be special problems not so far considered?

EXERCISE

3.1 Construct a preliminary block flow diagram for benzene hydrogenation, tabulating important features of the reaction and product separation stages and writing a process outline.

4. FLOWSHEET CONSTRUCTION

Even the first attempt at setting out the process scheme in its entirety —flowsheet construction — may bring to light particular difficulties whose solution decisively affects major questions of choice between alternative plant designs. This is illustrated by further consideration of the case of benzene hydrogenation to cyclohexane and by the flowsheeting of a process for dimethylterephthalate production.

After it has been established that design objectives are achievable by means expressed in an outline process description and block flow diagram, the next step is to select plant items and equipment capable of performing the desired operations. Mass and energy balances are now required; these, together with a knowledge of any relevant chemical kinetic data and mass and energy transfer rates, allow equipment to be sized, design assumptions verified and flowsheets developed. Standard texts and taught courses are available for all these steps.

The following case studies illustrate aspects of flowsheet development which are perhaps less well documented. In Study 4.1 we continue the development of the cyclohexane process, where we find that choice of reactor system is determined by considerations of heat removal. In Study 4.2 the reaction is the esterification of terephthalic acid (TA) by methanol; the process operates at high pressure and the difficulty of charging or feeding the solid TA to the reactor dominates the design choices.

CASE STUDY 4.1 — BENZENE HYDROGENATION

DESIGN OBJECTIVES AND CONSTRAINTS

Figure A3.3 given in Appendix II as the solution to Exercise 3.1 in the previous chapter is the block flow diagram for the proposed process whose objectives require benzene conversion to be at least 99.99%, with indicated maximum temperature of 175–200°C. Excess hydrogen is necessary to ensure this conversion. The reactor must be designed to withstand 40 bar to contain the requisite hydrogen gas and saturated liquid (mainly cyclohexane) at the elevated tempera-

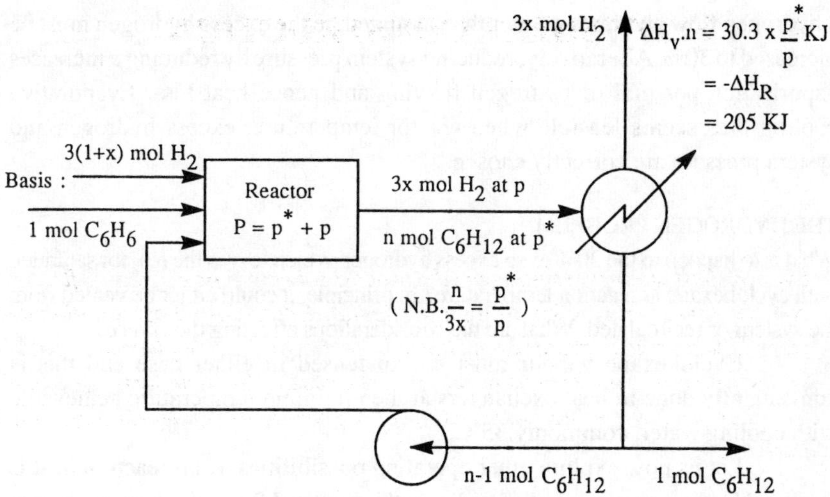

Figure 4.1 Preliminary simplified mass and energy balance around a benzene hydrogenation reactor.

ture. Heat removal is a special problem, since the reaction is exothermic and reversible with a heat of reaction of –205 kJ/mol, sufficient to vaporise 7 mol/mol product. How is this heat to be removed?

THE HEAT REMOVAL PROBLEM

Excess hydrogen is used and flows from the reactor saturated with hydrocarbon vapour at reaction temperature. The evaporation of hydrocarbon removes latent heat, so cooling the reactor. Can this maintain the desired temperature?

Trial mass and energy balances around the reactor will clarify this question. Figure 4.1 provides an approximate basis for making such trials at assumed values of *x*, the percent excess of hydrogen, *p* its partial pressure and p^* the vapour pressure of product cyclohexane at reactor temperature. If we assume provisionally that the hydrogen feed is in 5 mol% excess over the three mol required for one mol of product, then 3 × 0.05 = 0.15 mol leave the reactor with each mol of cyclohexane produced. This is saturated with cyclohexane at, say, 200°C when p^* is 13.25 bar. If the system pressure is to be 15 bar, *p* is 1.75 bar; then the cyclohexane evaporation rate is 13.25 × 0.15/1.75 mol/mol product and the latent heat extraction rate is 1.14 × 30.3 = 35 kJ/mol of product. The heat of reaction is –205 kJ/mol. To remove all this by evaporation evidently requires

a hydrogen flow six times greater than assumed, ie the excess hydrogen must be increased to 30%. Alteratively, reducing system pressure by reducing *p* increases vaporisation per mol of hydrogen flowing and hence heat loss. Evaporative cooling thus seems feasible when reactor temperature, excess hydrogen and system pressure are correctly chosen.

THE HYDROGEN PROBLEM

What is to happen to the 30% or so excess hydrogen which leaves the reactor saturated with cyclohexane at reaction temperature? In principle, it could either be vented from the system or recirculated. What are the considerations affecting the choice?

Cyclohexane vapour must be condensed in either case and this is conveniently done in heat exchangers at the minimum temperature achievable with cooling water, commonly 35°C.

Let us now explore other operating possibilities. If all reaction heat is removed by evaporative cooling, what is the required flow of excess hydrogen per mol of benzene feed when reactor temperature is 175°C at a system pressure of 15 bar? If this excess hydrogen is cooled to 35°C, what is its residual cyclohexane content? What consequences follow and what influence have these on the design?

A heat balance around the reactor can be obtained under a variety of operating conditions: 200°C reaction temperature, 15 bar system pressure and 30% excess hydrogen as in the previous discussion; 175°C, 15 bar and 61% excess hydrogen in the present example. In all cases, two consequences appear:
(1) The required rate of evaporation would exceed the rate of formation of product;
(2) Venting hydrogen leads to product loss.

In the first case above, condensing to a final temperature of 35°C when p is 0.2 bar, and venting all the excess at atmospheric pressure, we find a loss of 0.20 mol cyclohexane/0.80 mol hydrogen or 0.23 mol/mol product. In the further example, 0.45 mol/mol product is lost. What are the implications for the design?

In the first place, regarding Consequence (1) above, it is clear that cyclohexane must be condensed from the hydrogen leaving the reactor and the condensate must be returned as reflux to maintain reaction volume; if all the heat of reaction is to be rejected to a condenser, the rate of reflux or liquid recycle must substantially exceed that of product offtake. Consequence (2) leads to two alternatives: the excess hydrogen might be vented, in which case the accompanying product loss would be intolerable unless some cyclohexane absorption system followed the reflux condenser; alternatively, the hydrogen bearing its residual cyclohexane might be recycled, when such an absorption system is unnecessary. What are the relative advantages of these alternatives?

Recycling allows virtually complete utilisation of hydrogen, whatever its excess as supplied to the reactor; the resulting flexibility in the amount of excess assists temperature control and assures complete reaction. Gas recycling is, however, costly in energy and equipment. Venting uses no energy or expensive recirculation equipment but entails energy and equipment costs for absorption and stripping. Recycling is chosen.

What if the hydrogen feed is impure? Hydrogen produced by natural gas reforming and separation may contain traces of methane, carbon monoxide or carbon dioxide. If the impurities poison the catalyst, they must be removed; if non-reactive, they accumulate in the recycle and must be purged. In practice, methane and other inerts must be purged from benzene hydrogenation plant; traces of impurities are inevitably carried in with the reactants or else formed by secondary reaction. The ratio of purge to hydrogen feed determines purge — and recycle — composition. Lowering the ratio raises the impurity content; this reduces product loss but also lowers hydrogen partial pressure in the reactor, hence depressing equilibrium conversion. A balance must be struck.

KINETICS

The reaction rate equation must be determined experimentally, and unlike equilibrium conversion, cannot be calculated from thermodynamic data. A literature search for experimental information on benzene hydrogenation reveals that a nickel or precious metal catalyst has been used. A full design study would evaluate all catalysts, but in the present study we select Raney nickel, using information from existing plant which indicates the charge required. Raney nickel is liable to a slow deterioration in performance, from poisoning and mechanical degradation. What if too little or too much is charged?

Equilibrium data set an upper limit to the extent of reaction (which may not, of course, in practice be reached). Equilibrium conversion is a function of the original composition and the final temperature and pressure. If these are fixed by operating conditions, charging 'too much' catalyst costs money but has no other deleterious effect. Since reliable product quality is here a prime design objective with no opportunity for subsequent correction, a generous catalyst charge is indicated.

How can the catalyst and reactant flows be disposed within the reactor?

REACTOR SELECTION

Raney nickel is a particulate solid. With three phases accordingly present in the reactor, there are problems of distribution and contact. A static catalyst bed is here preferred to a mobile one, because catalyst attrition and other complications of particle motion are thus avoided, and since there are no problems of convective heat transfer from the dense phase when heat loss occurs through vaporisation of boiling liquid.

The catalyst pellets are supported on a perforated grid above graded inert packing layers. Liquid flow may be either upward, in which case the bed is permanently flooded, or downward, when careful attention to distribution and rate are necessary. With a flooded bed, hydrogen flow must be upwards. With a wetted (trickle) bed, gas flow may be co-currently downwards or counter-currently upwards. Which arrangement is preferable in the present case?

Current opinion favours a flooded bed, on the grounds of simplicity and flexibility. Hydrogen flow must then be upwards, distributed via a perforated feed pipe, known as a sparger, with an appreciable pressure drop resulting from small perforation size.

There is now sufficient information to draw a preliminary flow sheet covering the reaction stage.

FLOWSHEET DRAFTING

Flowsheets use standard symbols to represent plant items, showing these in elevation, approximately in their proposed spatial relation, with correct upward and downward flows. Main reactant flows are from left to right, recycles excepted. Such flowsheets do not usually fit A-standard paper conveniently and multiple sheets are advisable. For ease of reading and to avoid possible ambiguity, unique plant item numbers are preferred with letter prefixes indicating the type of equipment. Numbers and brief item titles are shown on the flowsheet. Mass balances and similar tabulated data are conveniently shown on separate typed sheets. Figure 1.1 in Reference 4.1 is a good example.

We now have to address the following tasks:

(a) Complete Figure 4.2 (pages 40 – 41) — 'Benzene hydrogenation: skeleton flowsheet' — to show the main features of the reaction stage. We assume that hydrogen is available already compressed and that cooling water is available at adequate pressure.

(b) Compile the corresponding preliminary plant item list, giving the title and brief functional description of each item.

At this stage omit arrangements for charging, discharging, activating or re-activating catalyst, purifying or pre-treating feeds, and details of test or storage tanks. On the complete flowsheet, all these items will of course be shown. Table 4.1 is the preliminary plant item list.

SOME PROCESS CONTROL ESSENTIALS

These should be accommodated while we proceed with task (a), flowsheet preparation. Various operating constraints have already been identified at different process stages. Reactor outflow is at 150 – 200°C, but product in storage, hydrogen recirculation and purge are to be as close to ambient temperature as practicable. Reactor pressure is in the range 10 – 40 bar. Excess hydrogen is required. Purged inerts and hydrogen are vented to a low pressure main. Product is stored at atmospheric pressure. Flowsheets should show the principal controls required to achieve these objectives. Is there a simple procedure for identifying these?

Consider first what controllable variables determine the desired mass flows; next, consider those affecting system pressure and finally those governing operating temperature. Determine at what locations (and by what means) the operating variables are to be sensed and where the appropriate controlling

TABLE 4.1
Preliminary plant item list for benzene hydrogenation

Plant item	Title and function
T101	Benzene feed tank
P102	Benzene feed pump
C201	Hydrogen recycle circulator
R301	Reactor (a cylindrical vessel)
E302-3-4	Condensers to remove cyclohexane from recycle hydrogen
E305	Cyclohexane product cooler
V401	Separator to remove droplets of condensed cyclohexane after hydrogen expansion through the control valve
V402	Separator to remove hydrogen evolved when total pressure falls as the recycle stream passes through the control valve
T601	Cyclohexane product tank

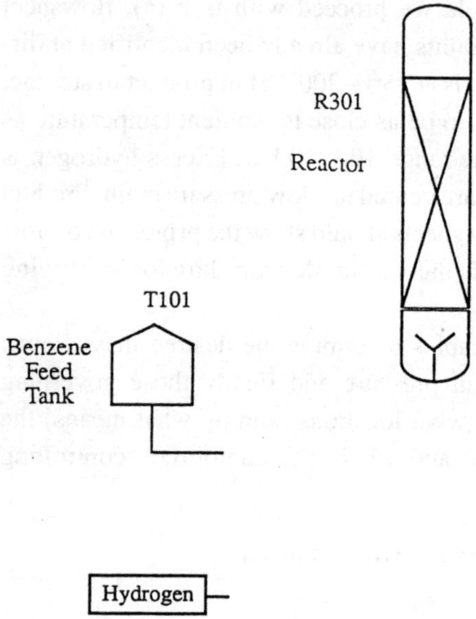

Figure 4.2 Benzene hydrogenation: skeleton flowsheet.

actions are to be applied. Bear in mind that to maintain reactor pressure one should restrict the outlet; to govern flow, control the inlet. The flowsheet in its present state of development is shown in Figure 4.3 (see pages 42 – 43).

The four main process controls are shown on this preliminary flowsheet. Their functions are explained in Table 4.2.

HOURLY AND YEARLY PRODUCTION RATES

Market surveys generally conclude that a plant is to produce a stated quantity of product in a year, whereas designers' flowsheet data are generally expressed in hourly rates. For continuous plant, it is usual to assume a production year of 8,000 hours. Table 4.3 (see page 44) shows how this operating time is arrived at for a particular case.

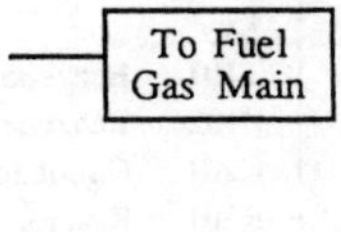

DESIGN PROCEDURES ARE ESSENTIALLY ITERATIVE

In Case Studies 3.2 and 4.1 we examined the cyclohexane process starting from a consideration of the separation of the product from the reaction mixture. Would the outcome have been the same if reaction had been considered before product separation? Probably yes, although it is unlikely that appropriate reaction conditions could have been specified in ignorance of separation problems or market demands. But in any event, as pointed out in Chapter 1, design is an iterative procedure. After specifying reaction conditions it is advisable in any case to re-examine the separation stage. In the present process there is a parallel reaction producing the cyclohexane isomer methylcyclopentane. How much is produced? Is it separable? Is it acceptable in the product? These and similar questions must be addressed during the design cycle.

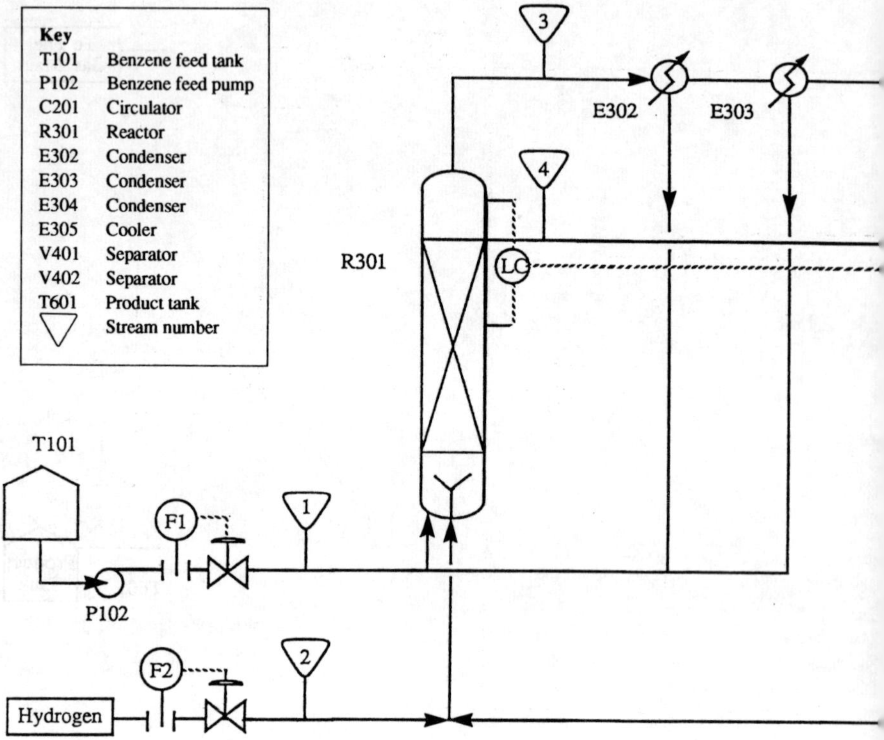

Figure 4.3 Benzene hydrogenation: preliminary flowsheet.

TABLE 4.2
Benzene hydrogenation control loop schedule

F1	Controls benzene flow to the reactor by throttling feed pump outlet flow.
F2	Controls hydrogen flow by throttling feed line (not applicable for the case of a reciprocating feed compressor).
LC	Liquid level control on the reactor to ensure catalyst is always flooded. The throttle valve is placed in the cool stream after E305 to minimise flashing with consequent valve wear.
PC	Controls pressure of the reactor section by throttling the gas vent in the coolest position.

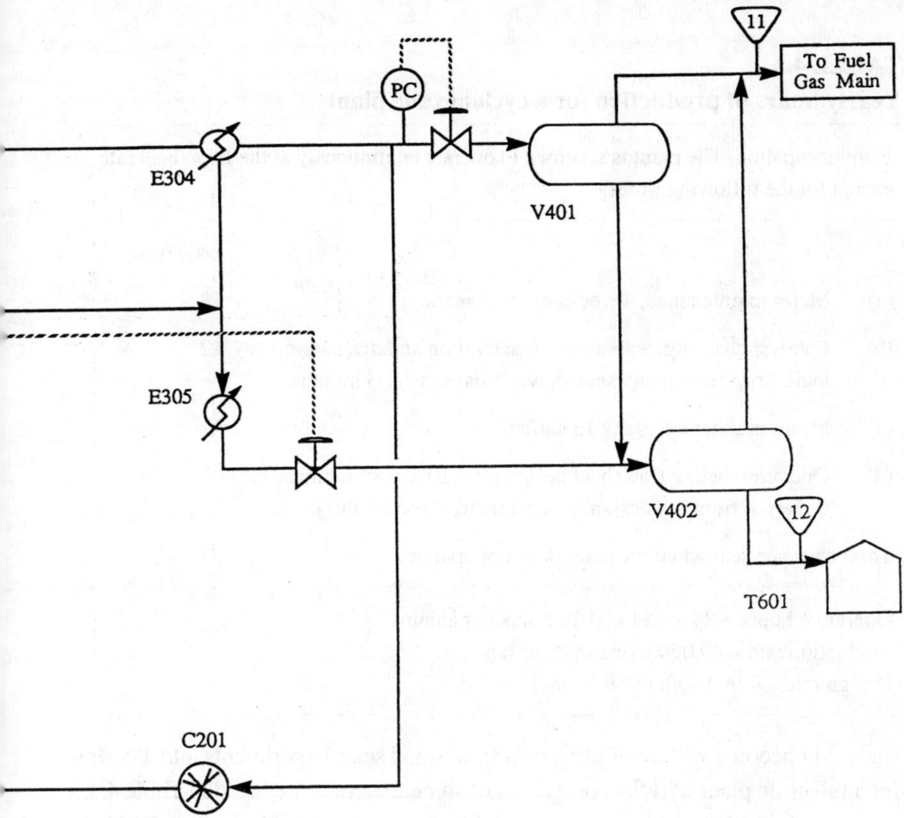

CASE STUDY 4.2 — ESTERIFYING TEREPHTHALIC ACID WITH METHANOL

PROCESS PRINCIPLES

The process objective is to produce dimethylterephthalate (DMT), an intermediate for polyester fibre manufacture. Two mol methanol react with 1 mol terephthalic acid (TA) forming DMT and water in mildly exothermic reversible esterification. The market requires exceptionally pure DMT and this can be achieved by recrystallising the product from methanol and distilling it. The main constraint on the reaction is to secure over 95% conversion, producing a solution from which DMT can be crystallised. There is literature information on the thermochemistry and equilibrium of the reactions but the present

TABLE 4.3
Yearly hours of production for a cyclohexane plant

Plant occupation. The plant is assumed to operate continuously at the flowsheet rate except for the following times:

		Days/year
(a)	Major maintenance, 4 weeks every 2 years:	14
(b)	Catalyst charging, activation, re-activation and discharge, including start-up and shut-down. 3 days every 3 months:	12
(c)	Minor maintenance, say 15 shifts:	5
(d)	Operation below flowsheet design rate (10% is sometimes deducted from the design hourly rate to allow for this):	Nil
Total days per year when the plant does not operate:		31

Operating hours = 24 × 334 = 8016 hours per annum
Production rate = 40,000 t/year cyclohexane
Design rate = 4,990 kg/h or 59.5 kmol/h

study is concerned with translating data from small scale experiments into a design for a full scale plant. Physical properties of some materials are listed in Table 4.4.

Reference 4.2 reports a satisfactory process as follows: 'DMT is prepared from TA with methanol at 50 psi and 110°C in the presence of sulphuric acid. The ratio of methanol and TA should be 15 – 20:1 and the amount of catalyst about three times the TA. The reaction is complete in 0.5 h.'*

Before we can convert this information into a preliminary block flow diagram we must note the objectives, constraints and special problems encountered, from receiving reactants to reactor outflow, ie entry to the separation stage.

* The reader may find it surprising that the precision of statements made in the patent literature is less than would be expected in a good textbook or research paper. This is because patents are concerned only with facts and not with their presentation. Even less are ideas or theories of any importance in such documents, which serve merely to claim priority for accomplishments.

TABLE 4.4
DMT production. Extracts from the data book.

Terephthalic acid (TA, 1,4-benzenedicarboxylic acid)

CA Registry Number:	100–21–0			
Formula:	$C_6H_4(COOH)_2$			
Molecular weight:	166.1			
Form:	Crystalline needles			
Phase behaviour:	Terephthalic acid sublimes without melting above 300°C			
Solubility data:				
Temp (°C)	25	120	160	200
g/100g water	0.0019	0.07	0.38	1.8
g/100g methanol	0.10		2.9	

Dimethylterephthalate (DMT, 1,4-benzenedicarboxylic acid dimethyl ester)

CA Registry Number:	120 – 61 – 6	
Formula:	$C_6H_4(COOCH_3)_2$	
Molecular weight:	194.6	
Form:	Crystalline needles	
Melting point:	141 – 142 °C	
Boiling point:	288°C at 760 mmHg pressure	
Solubility data:		
Temp (°C)	25	60
g/100g water	Negligible	Negligible
g/100g methanol	1.0	6.0

SPECIAL PROBLEMS

(1) Solubility and phase conditions. In what phase do the reactants combine?

In the previously considered case of benzene hydrogenation the reactor contained hydrogen and organics in the vapour phase, together with liquid organics. Hydrogen has some solubility in the liquid phase so that reaction occurs there in contact with catalyst, although it can also occur in the vapour phase. For the present process, the data book shows that at ambient conditions methanol is liquid and that at reaction temperature its vapour pressure is substantial. TA is a solid which sublimes at 300°C. It is only sparingly soluble in methanol. According to Reference 4.2, one gram of TA reacts with 5 – 6.7g of methanol at 105 – 110°C whereas, even at 160°C, 33.5g methanol is required for complete solution. At the start of the reaction, therefore, the fraction of dissolved TA is not greater than 0.2.

(2) Charging reactants. If all the TA is added to the methanol at the start of the reaction, then the reactor will contain a slurry and TA dissolves as the reaction progresses. The batch experiments reported in the reference must have been performed in this way (but since an autoclave was used, the physical condition inside could not be seen). Can TA/methanol slurry containing some 50 – 70% solids be handled satisfactorily in a continuous full scale plant? What are the possible alternatives to be investigated?

(a) Is it possible to prepare a cold slurry, preheat it and feed it to a hot reactor?

(b) Powder cannot be preheated; but can cold powder and hot methanol be fed separately to a hot reactor?

(c) Is there any other alternative?

(3) Operating pressure. Methanol vapour pressure at reaction temperature is over 4 bar. Experiments show that TA/methanol slurries are tractable provided they are continuously agitated. Pumping slurry to a hot reactor, while not impracticable, poses mechanical problems. The alternative of charging cold powder into a hot pressurised vessel is not practicable with a conventional rotary star feeder. A pressurised lock hopper system is required, with problems of purging methanol vapour from the feeder and inerts from the reactor. Alternative (b) should be avoided if possible, particularly if accurate weight ratios are required.

THE PREFERRED PROCESS

In this instance, therefore, a batch process is preferred to a continuous one, because this avoids the problem of continuously feeding powdered solid to a

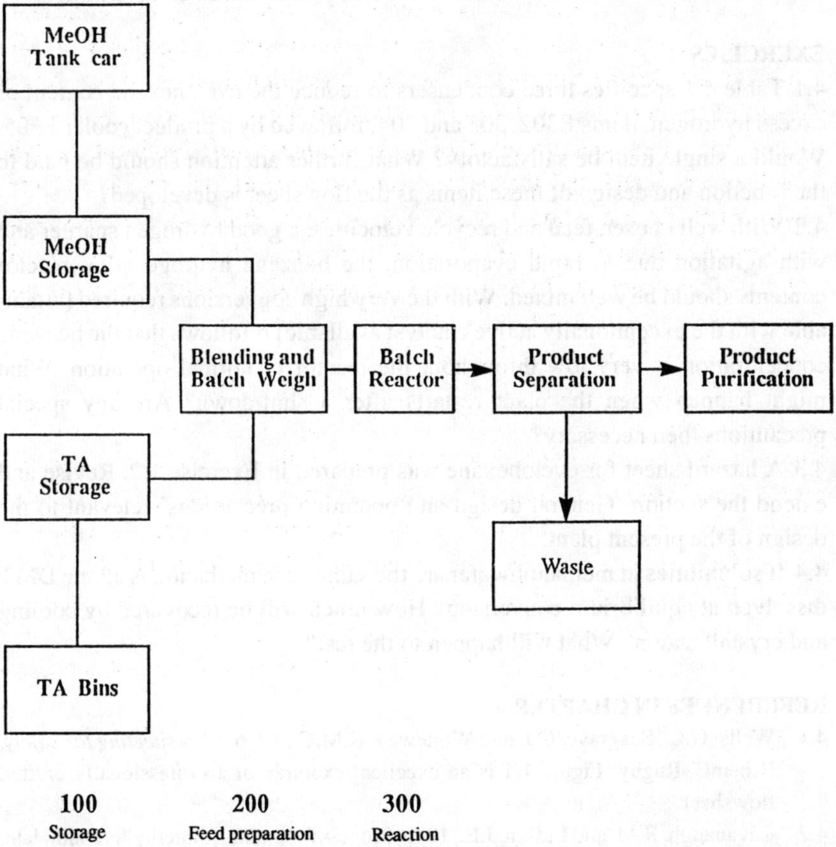

Figure 4.4 Preliminary block flow diagram for batch production of dimethyl terephthalate.

pressurised vessel. This vessel then performs more than one function, as described in the following outline process description accompanying the preliminary flowsheet, Figure 4.4: 'Methanol is received from tank cars into storage block 100, tested and blended with recovered methanol in block 200 and charged cold to the reaction stage in block 300. TA is received in bins, stored in area 100 and transferred to area 200 where batch charges are weighed into the reaction stage. Here the methanol/TA slurry is well agitated, heated to 105 – 110°C and maintained at this temperature until required conversion is achieved.' How DMT is separated from the reactor contents downstream is still to be determined.

EXERCISES

4.1 Table 4.1 specifies three condensers to reduce the cyclohexane content of excess hydrogen, items E302, 303 and 304, followed by a product cooler E305. Would a single item be satisfactory? What further attention should be paid to the function and design of these items as the flowsheet is developed?

4.2 With well chosen feed and recycle velocities, a good hydrogen sparger and with agitation due to rapid evaporation, the benzene hydrogenation reactor contents should be well mixed. With the very high conversions required (attainable with the exceptionally active catalyst available) it follows that the benzene concentration is very low throughout the reactor in normal operation. What might happen when the plant restarts after a shut-down? Are any special precautions then necessary?

4.3 A hazard sheet for cyclohexane was prepared in Exercise 1.2. Revise and extend the section 'General design and operating precautions' relevant to the design of the present plant.

4.4 If solubilities in methanol/water are the same as in methanol, is all the DMT dissolved at equilibrium conversion? How much will be recovered by cooling and crystallisation? What will happen to the rest?

REFERENCES IN CHAPTER 4

4.1 Wells, G.L., Seagrave, C.J. and Whiteway, R.M.C., 1976, *Flowsheeting for safety*, IChemE, Rugby. Figure 1.1 is an excellent example of a professionally drafted flowsheet.

4.2 Cavanaugh, R.M. and Lufkin, J.E., 1949, Process for making dimethylterephthalate, US Patent 2,459,014 of 11th January 1949, reported in *Chemical Abstracts 1949 43, 3039g*.

5. THE EXPLORATION OF ALTERNATIVES

Process design or synthesis is essentially the selection and ordering of possible process operations. Processing steps may often be combined in several different, but functionally equivalent, sequences. These must be examined to disclose the special problems and advantages of each, allowing the designer to identify the best and cheapest schemes for development.

PROCESSING SEQUENCES

An all-pervading feature of chemical process design work is multiplicity of choice: choice of product to meet a market demand, of reactants to convert to a given product and of the sequence of operations within a process. A basic generic flowsheet follows the virtually invariable sequence — feed preparation, reaction, separation, product purification — but within the 'blocks' of this sequence there are usually significant opportunities for alternative arrangement.

In a few particular circumstances the best arrangement of process steps is uniquely determined by physical or chemical considerations. These instances where the designer has no real choice are generally well known and readily recognised. Thus, heat transfer operations generally show the greatest economy of heat recovery and usage when ordered according to the principle 'hottest hot against hottest cold', and by applying pinch design methods (References 1, 2 and 3). Rules for ordering distillation sequences for maximum efficiency are also well documented. Similarly, reaction sequences may be dictated by chemical considerations. Thus 1-chloro-3-nitrobenzene is produced by nitrating benzene first and then chlorinating the nitrobenzene. The order of these steps may not be reversed; chlorination followed by nitration produces 1-chloro-2-nitrobenzene and the 1,4 isomer. In general, however, no single consideration or principle decides the optimum order of process steps. The following case studies illustrate the procedures which must then be followed.

In Case Study 5 we discuss an air oxidation process for 'cyclohexanol' based on the reaction scheme of the ill-fated Flixborough plant, where isolating

TABLE 5.1
Product distribution in the air oxidation of liquid cyclohexane

Component	Weight percentage
Cyclohexane	88
Cyclohexanol	5
Cyclohexyl esters	6
Water	1

product from the reactor outflow requires a second reaction and at least two separations. No readily deducible rule or principle determines the best order for these steps, so all possibilities must be examined in turn. Systematic investigation of alternative sequences shows that only one achieves the design objectives with the theoretical minimum number of operations. This is the preferred process.

CASE STUDY 5 — ISOLATING CYCLOHEXANOL

There is a market for cyclohexanol/cyclohexanone mixture as an intermediate for nylon manufacture. In this study the marketable mixture is called cyclohexanol for simplicity. Liquid phase air oxidation of cyclohexane produces a reactor outflow with the composition given in Table 5.1.

One design objective is to maximise the overall yield of desired product by utilising both free and combined cyclohexanol components of the stream leaving the reactor. The esterified cyclohexanol is released by hydrolysing the esters with dilute caustic soda solution, producing carboxylic acid sodium salts as waste products. Table 5.2 gives some properties of these materials.

How many components are present after hydrolysis and what is the least number of operations required to separate them?

Components present after hydrolysis, whenever it is performed, are as follows:

- unreacted cyclohexane for recycle;
- cyclohexanol;

TABLE 5.2
Cyclohexanol production: extracts from the data book

	Cyclohexane	Cyclohexanol	Acids	Esters	Salts
BPt, °C	69.8	161.1*	—	>200	—
MPt, °C	6.5	24.8	—	0	—
Solubility:					
In water	No	All propns.	Sp	Negligible	Yes
In h'bons	Yes	Yes	No	Yes	No

* Forms azeotrope bp 97.8/760 mm c.20% cyclohexanol:80% water

- sodium salts;
- water introduced with the caustic solution.

The number of components is effectively three, since the salts dissolve in water; not less than three operations are accordingly required for their separation.

What properties of these components are relevant to their separation? What are the corresponding constraints and operations?

POSSIBLE OPERATIONS

Properties, constraints and operations to consider include:

- miscibility, for phase separation. Cyclohexane and water are immiscible. Cyclohexanol is miscible with both, but its partition into the organic phase will be enhanced if the aqueous phase contains dissolved salts.
- volatility, for distillation or evaporation. Cyclohexane and water are more volatile than cyclohexanol, but the three components form two immiscible layers. What happens if the three components are distilled together?

If distilled together, the partial pressures of the two phases are additive, so that at 760 mmHg the bubble point is 70°C and the distillate contains both. The bottoms may be either aqueous or organic, according to feed composition. Salts dissolved in an aqueous layer depress the vapour pressure. Evaporating to dryness is thus impracticable, because readily available heating media are not stable at the high temperature required. Operations to be investigated are therefore distillation, phase separation and the necessary hydrolysis reaction.

What sequences are to be considered for these operations?

POSSIBLE SEQUENCES

Three operations can be arranged in six possible ways, viz:

(1) Distillation — hydrolysis — phase separation;
(2) Distillation — phase separation — hydrolysis;
(3) Phase separation — distillation — hydrolysis;
(4) Phase separation — hydrolysis — distillation;
(5) Hydrolysis — distillation — phase separation;
(6) Hydrolysis — phase separation — distillation.

Our procedure must now be to attempt to construct block flow diagrams for each of the foregoing sequences in turn. When particular problems arise we proceed to the next sequence, finally selecting a preferred scheme with fewest major problems, which fulfils the design objectives and which can be developed to the flowsheet stage.

Sequences (1) and (2) — These satisfy familiar guidelines for distillation, removing first the component with the greatest mass and highest volatility, viz cyclohexane. Hydrolysis is not yet performed, so components present are cyclohexanol, esters and a trace of water. These form virtually a single organic layer, so phase separation before hydrolysis, indicated in Sequence (2), would be ineffective. In Sequence (1), hydrolysis follows without difficulty, but can cyclohexanol be recovered from the residual aqueous salts solution?

Experimental work is required to find whether one or two layers form. While cyclohexanol is partially miscible with water, sodium salts in the aqueous layer reduce miscibility. Thus phase separation cannot be used to remove all the cyclohexane; Sequence (1), like Sequence (2), would not be effective. Other sequences are accordingly examined.

Sequences (3) and (4) — As in Sequence (2), phase separation before hydrolysis is not effective.

Sequences (5) and (6) — Hydrolysis results in two phases, cyclohexane/cyclohexanol and salts solution containing traces of cyclohexanol depending on its partition coefficient. Does pursuing Sequence (5) to the distillation stage disclose any potential difficulties?

Distilling both phases together resembles steam distillation; wet cyclohexane goes overhead, leaving cyclohexanol/water/salts as bottoms. This mixture will not phase-separate, so this sequence looks inappropriate. Sequence (6), hydrolysis followed by phase separation, yields an organic layer which can be distilled to give cyclohexane overhead with cyclohexanol product below, while the aqueous layer with its dissolved salts can be discarded. Any small cyclohexanol content might be recovered in a further operation, if economic.

Thus, Sequence 6 appears feasible, employs the minimum theoretical number of operations and is preferred for developing to the flowsheet stage. Assumptions about physical properties and behaviour are to be confirmed by experiment where necessary. Figure 5.1 (see pages 54 – 55) is a schematic flowsheet for this scheme.

Air and excess cyclohexane are fed to reaction stage 300. This is highly exothermic, so that nitrogen leaves the reactor at 130 – 150°C with cyclohexane vapour, which is condensed in block 400. Block 500 is an additional adsorption step before nitrogen is vented. Block 600 is the second reactor, for hydrolysis. After separation, the organic layer is distilled in block 700 to recover unreacted cyclohexane, while the mixed product is taken as bottoms.

EXERCISES

5.1 The reactor temperature used implies a substantial pressure of vapour in contact with the liquid. Does this call for any particular provision in the flowsheet?

5.2 Cyclohexanol has a reported melting point of 25°C. Could it be separated by chilling?

5.3 If you were buying the product produced according to Figure 5.1, what specification clauses would you pay particular attention to?

5.4 Is it realistic to consider incinerating the organic salts directly in aqueous solution?

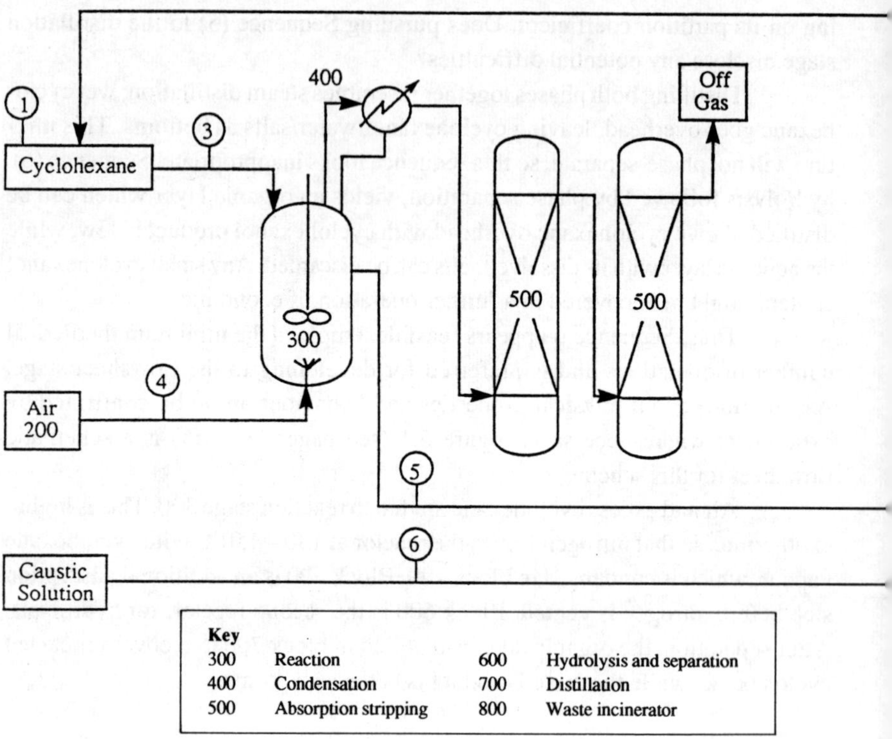

Figure 5.1 Schematic flowsheet for cyclohexanol production.

REFERENCES IN CHAPTER 5

5.1 Ponton, J.W. and Donaldson, R.A.B., 1974, A fast method for the synthesis of optimal heat exchanger networks, *Chem Eng Sci*, 29: 2375–77.

5.2 Townsend, D.W. and Linnhoff, B., 1981, in *Total energy design in process plants*, IChemE/SCI Symposium, 36–46.

5.3 Linnhoff, B. and Hindmarsh, E., 1982, in *Understanding process integration, IChemE Symposium Series No. 74*, 67–98.

5.4 *The Flixborough disaster*, 1975, Department of Employment, HMSO.

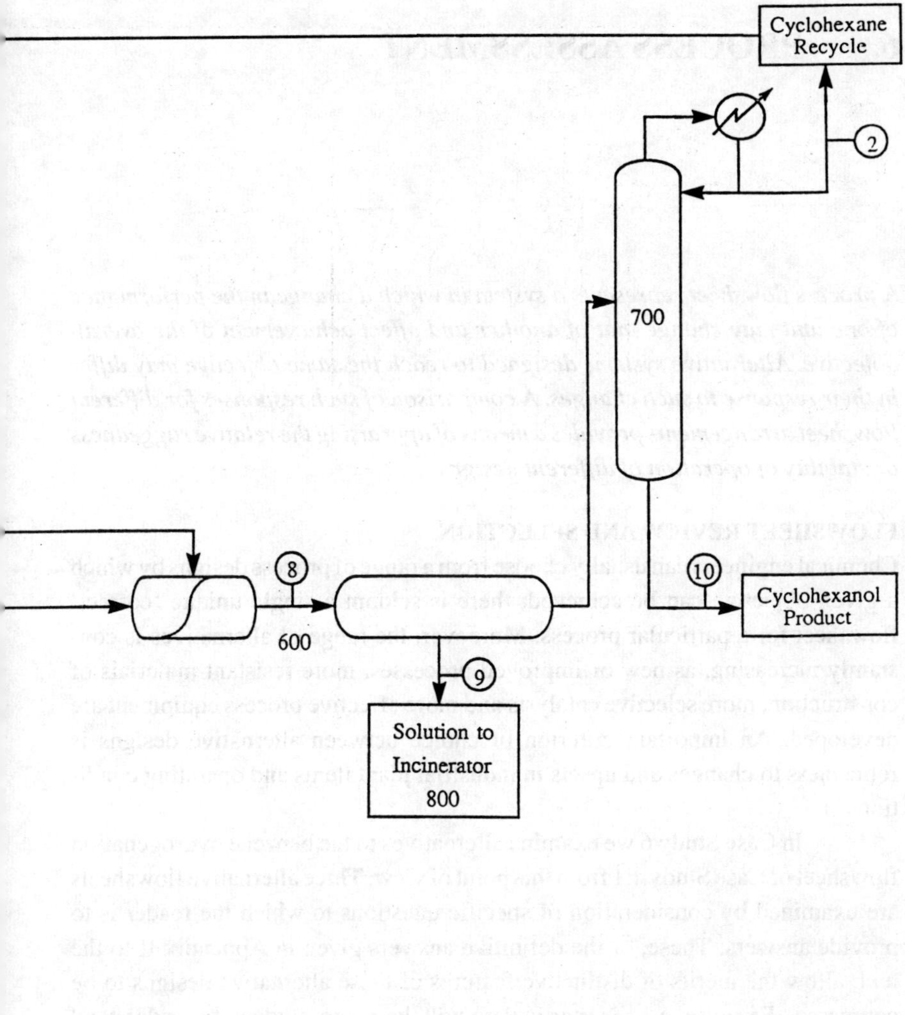
Cyclohexane
Recycle
2
700
8
10
Cyclohexanol
Product
600
9
Solution to
Incinerator
800

6. PROCESS ASSESSMENT

A process flowsheet represents a system in which a change in the performance of one unit may change that of another and affect achievement of the overall objective. Alternative systems designed to reach the same objective may differ in their response to such changes. A comparison of such responses for different flowsheet arrangements provides a means of appraising the relative ruggedness or stability of operation of different designs.

FLOWSHEET REVIEW AND SELECTION

Chemical engineers can usually choose from a range of process designs by which a given objective can be achieved; there is seldom a single unique 'correct' flowsheet for a particular process. Moreover, the range of alternatives is constantly increasing, as new or improved processes, more resistant materials of construction, more selective catalysts and more effective process equipment are developed. An important criterion of choice between alternative designs is robustness to changes and upsets in industrial plant items and operating conditions.

In Case Study 6 we examine alternatives to the benzene hydrogenation flowsheet of Case Study 4.1 from this point of view. Three alternative flowsheets are examined by consideration of specific questions to which the reader is to provide answers. These, or the definitive answers given in Appendix II to the text, allow the merits of distinctive features of these alternative designs to be compared. Exercise 6.1, in particular, will help the student to understand flowsheets and to appreciate the operating characteristics of the plant they represent; the other exercises give practice in identifying areas where local variations, such as pump failures, fouling of heat exchange surfaces, progressive formation of catalyst 'fines' and throughput changes, have a critical effect on overall performance. Such a comparative analysis of trial flowsheets would be used to generate further designs free of these faults.

CASE STUDY 6 — REVIEW OF CYCLOHEXANE FLOWSHEETS

THE DESIGN PROBLEM

The design objective is to achieve virtually complete benzene hydrogenation, despite the fact that the reaction is exothermic and that temperature rise reduces equilibrium conversion and thus yield. Either adiabatic or (more costly) cooled reactors may be selected, or some combination thereof. A design constraint on the reactor is the necessity of securing adequate contact between the liquid phase and both reactant gas and solid catalyst. These considerations lead to the elaboration of the simple catalyst bed shown in Figure 4.3, to a design in which there are three beds with inter-stage cooling, each operating adiabatically. The resultant flowsheet is Figure 6.1 (see pages 58 – 59).

Internally cooled catalyst beds are also feasible and Figure 6.2 (pages 60 – 61) represents such a system, while the design of Figure 6.3 (pages 62 – 63) uses both liquid phase and gas phase reactors.

Energy conservation is an important design consideration and this exothermic reaction offers possibilities of energy recovery by raising steam in waste heat boilers. The flowsheets show how this can be done in three different ways.

FLOWSHEET 6.1

Case Study 4.1 contains preliminary explanatory information and a preliminary plant item list. In Figure 6.1 the original process scheme has been developed into one employing a three-stage catalyst bed with inter-stage cooling. E1303 is now a waste heat boiler raising steam with feed water preheated in E1313.

Because boiler feed water is usually a warm mixture of returned condensate and treated water, E1314 is necessary to reduce the cyclohexane content of the recirculating hydrogen by further cooling with cold water.

EXERCISES ON FLOWSHEET 6.1

(1) What is the purpose of having three catalyst beds in the reactor?

(2) How is the reaction section maintained at operating pressure?

(3) The hydrogen feed contains a trace of inerts. Is the concentration of inerts in the recirculating hydrogen stream different from that of the feed? Will any such difference affect process operation?

(4) How can the rate of purge be varied while maintaining constant reactor pressure?

(5) E1303 and 1313 must be at a slightly lower pressure than R1301, yet no pump is shown to return condensate to the reactor. Is this feasible?

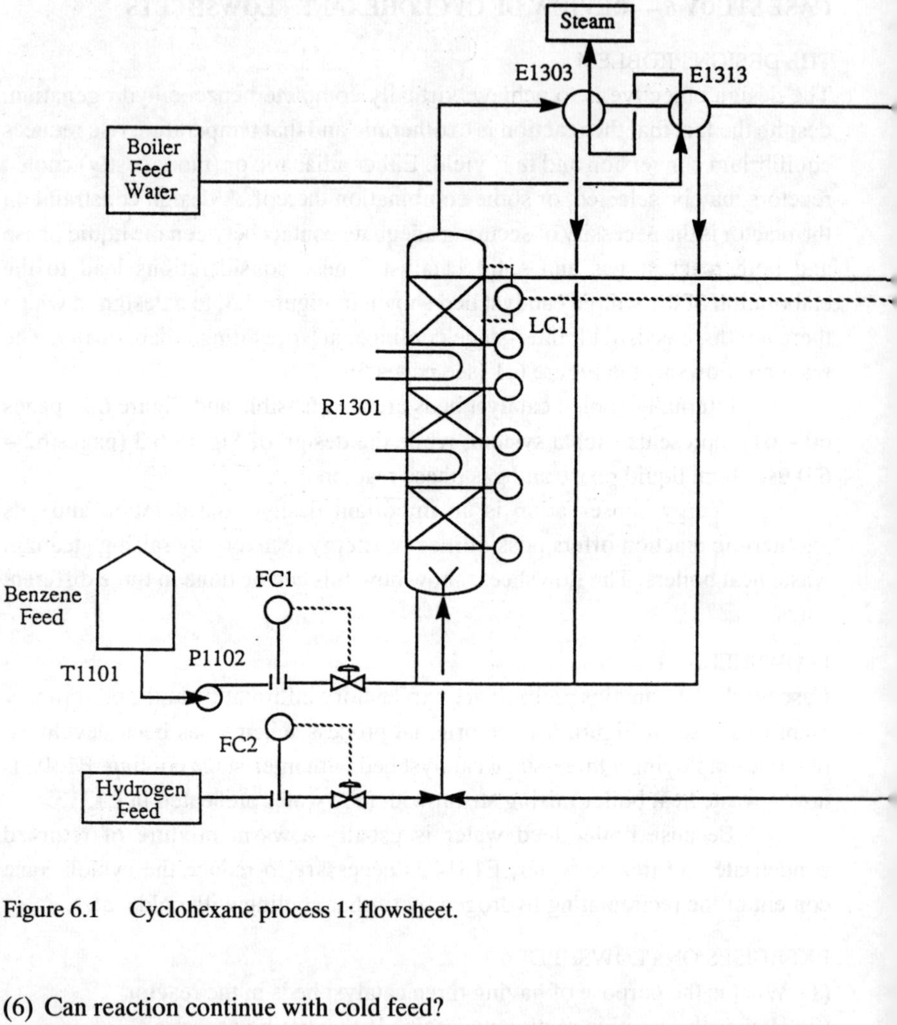

Figure 6.1 Cyclohexane process 1: flowsheet.

(6) Can reaction continue with cold feed?
(7) What improvements are advisable in the arrangements for flow into T1601?

FLOWSHEET 6.2

This is a simplified version of a process using Raney nickel suspended in the reactants.

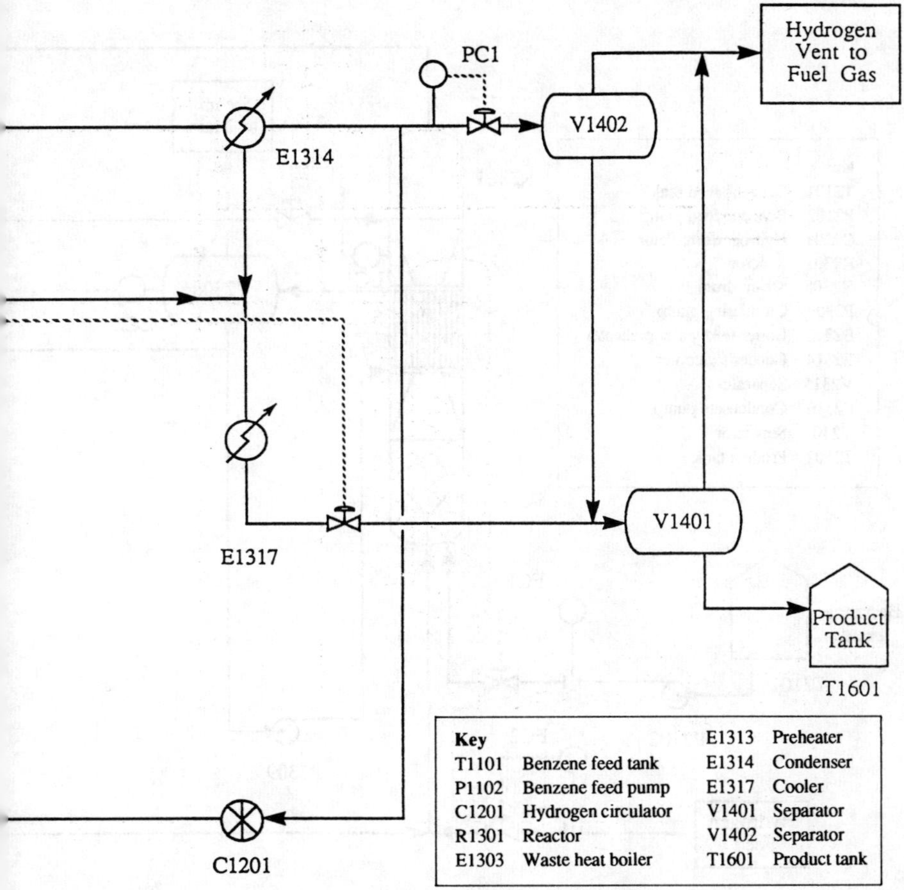

EXERCISES ON FLOWSHEET 6.2

(1a)How is reactor liquid level maintained? (b) How does this method compare in reliability with that of Process 1?

(2a)How is the heat of reaction removed? (b) What merits has this method compared with that of Process 1?

(3) How is reactor pressure maintained?

(4) Why is V2401 necessary?

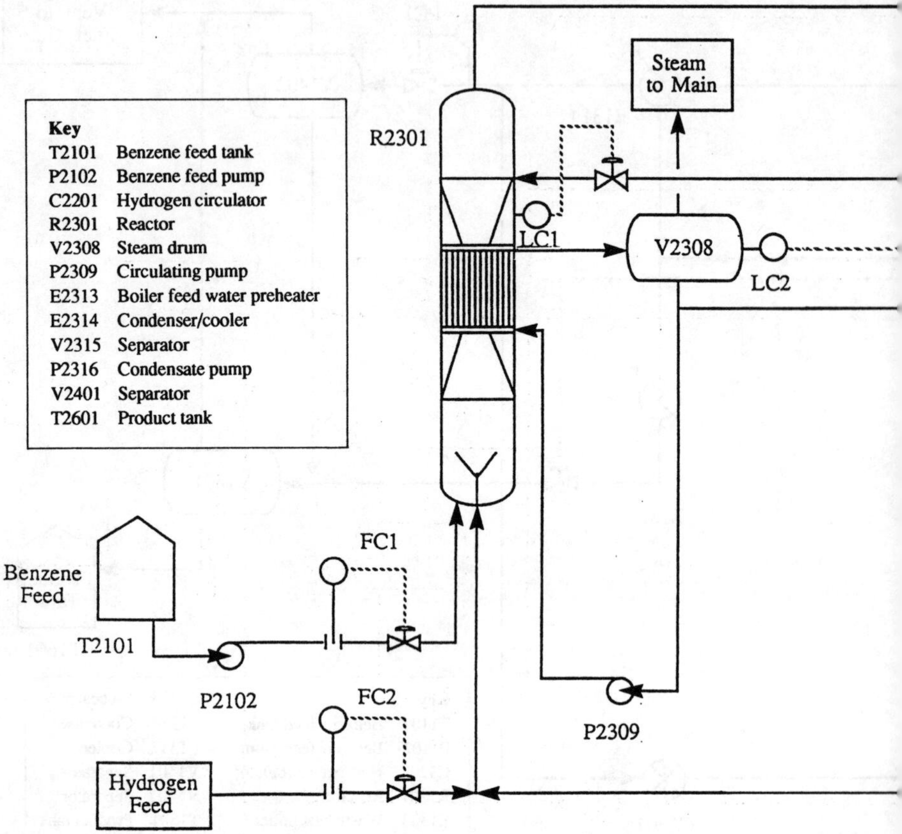

Figure 6.2 Cyclohexane process 2: flowsheet.

(5) What special problems of operation appear when this reaction system is started up?

(6) By what operating procedures and flowsheet modifications might any such difficulties be avoided?

(7) What control problems are associated with steam drum V2308?

(8) What are the principal shortcomings of the arrangements for venting or purging?

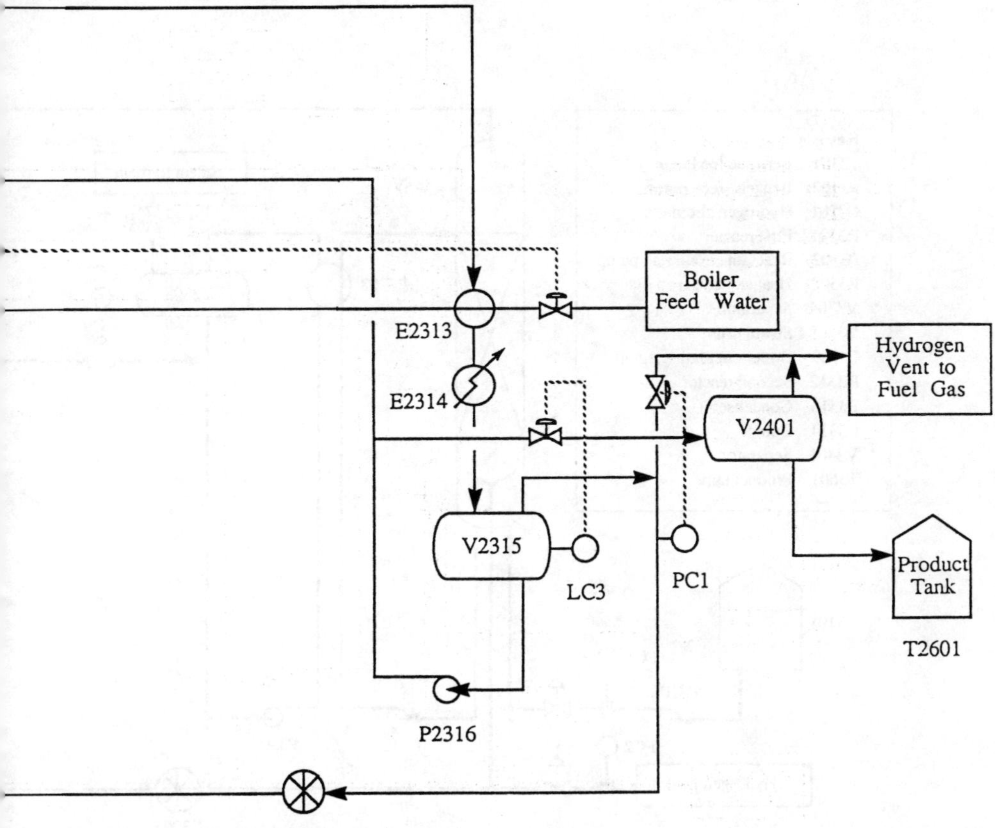

FLOWSHEET 6.3

Figure 6.3 is the simplified flowsheet of a process described in Reference 6.1. The catalyst in the first reactor is Raney nickel dispersed in reactants, while the second contains a fixed bed of alumina-supported nickel catalyst.

EXERCISES ON FLOWSHEET 6.3

(1) How is the liquid level maintained in the first reactor?

(2) How is the heat of reaction removed?

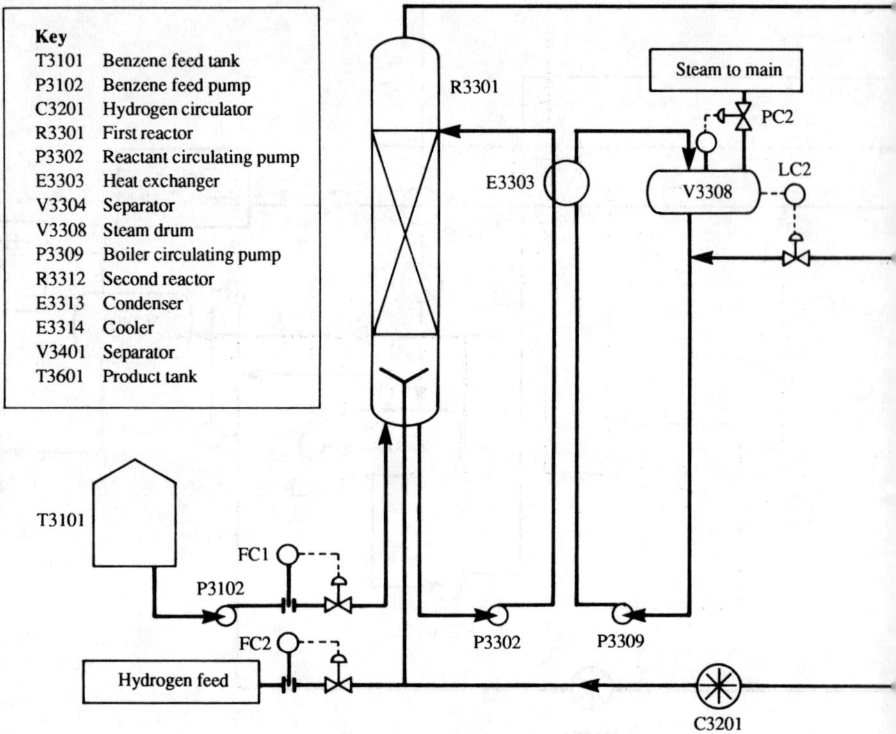

Figure 6.3 Cyclohexane process 3: flowsheet.

(3) What advantages or disadvantages has this scheme of heat removal?

(4) What difficulties do you foresee in the operation of E3303?

(5) What is the function of the second reactor? How will its duty vary with time and how can this variation be monitored?

(6) What is the function of V3401? What controls should be fitted to it?

(7) What special precautions are necessary in specifying the operating conditions of C3201?

(8) How is the plant kept at the required operating pressure?

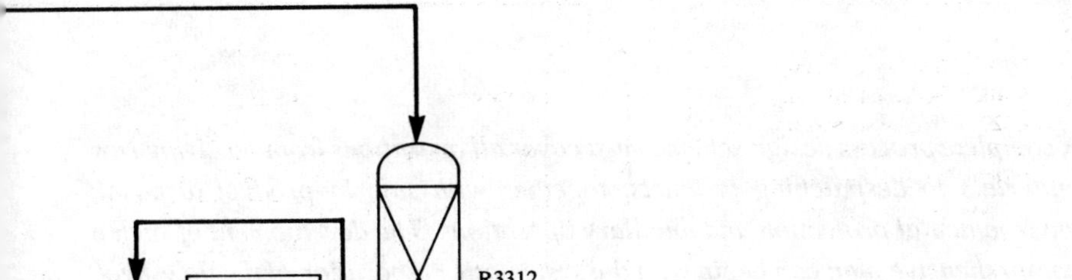

(9) If hydrogen is supplied from a reciprocating compressor, why is the arrangement of FC2 unsatisfactory as shown? Sketch an acceptable arrangement.

REFERENCES IN CHAPTER 6

6.1 Dufau, F.A., Eschard, F., Haddad, A.C. and Thonon, C.H., 1964, *Chem Eng Prog* 60 (9): 43–47.

6.2 *European Chemical News*, 19 April 1974.

7. STEPS TOWARDS COMPLETE PROCESS SPECIFICATION — ELABORATING THE FLOWSHEET

A complete process design scheme must cover all operations from receiving raw materials to despatching products, together with safe by-product disposal, environmental protection and ancillary operations. The development of such a comprehensive plan can begin with the systematic elaboration of the flowsheet.

THE DESIGN PACKAGE

A complete process plant is the outcome of a cooperative effort between people practising many different disciplines. The chemical engineering designer has the responsiblity of coordinating their activities. He must therefore provide sufficient documentation on 'his' design (which will itself be the product of much cooperative effort) to communicate his intentions to others without misunderstanding or the need for repeated requests for clarification. There is more to design than chemical engineering calculation.

Typically, the required documentation includes a process flowsheet, mass and energy balances, preliminary equipment data sheets, effluent treatment methods, process yields, process control methods, equipment layout and other information.

Good examples of a complete documentation package and of a process flowsheet, the most informative single item among these documents, are given respectively in References 7.1 and 7.2.

How can the designer check that such documents will fully and effectively communicate his intentions? One method is by following on the process flowsheet the flow of each material from reception to ultimate disposal, visualising the plant operating instructions which would be necessary to maintain these flows safely and checking that the equipment specified and its arrangement are appropriate. In the following case studies, a chemical engineer has prepared preliminary design proposals for a fructose syrup plant. His flowsheets for carbon treatment and ion exchange regeneration are now examined to see what further design provisions are required to enable the necessary process operations to be performed effectively with the plant and equipment specified in the preliminary flowsheet.

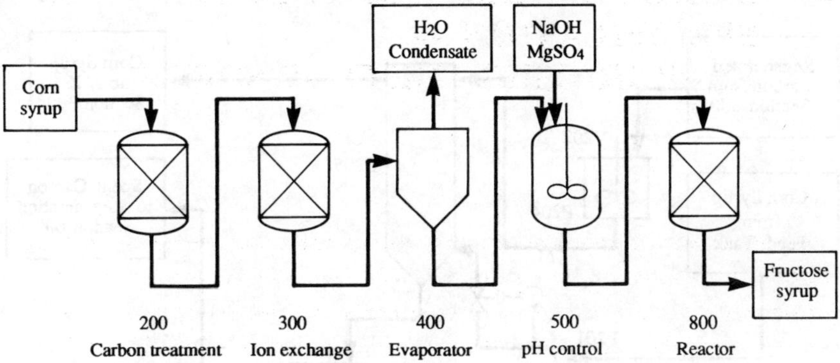

Figure 7.1 Block flow diagram for manufacture of fructose syrup.

CASE STUDY 7.1 — CARBON TREATMENT OF CORN SYRUP

DESIGN OBJECTIVES

Fructose syrup is prepared by isomerising the glucose component of corn syrup in an equilibrium reaction promoted by an immobilised enzyme. Proper feed preparation, which extends enzyme life, includes carbon treatment shown in block 200 of Figure 7.1, followed by ion exchange in block 300. For block 200 there is a choice between using granular carbon in fixed beds or pelleted carbon in a pulsed bed of proprietary design, from which spent carbon is drawn periodically for regeneration by controlled partial combustion. The latter option is selected.

PROCESS DESCRIPTION

What additional operations are necessary at the carbon treatment stage?

In addition to the means of treating the main process stream, the flowsheet should show means of withdrawing and recharging carbon to and from the regeneration section. An outline scheme is shown in Figure 7.2 (see page 66).

A slurry of carbon in syrup is withdrawn by gravity, and then agitated in tank V207 from which it is transferred to the regeneration section 600 to be shown on a separate flowsheet. Slurry transfer uses an air operated slurry pump, or similar device, shown as P209. Regenerated carbon is similarly recharged, as a water slurry. These arrangements lose syrup to the regeneration plant and dilute syrup in the process, and require further consideration. During regeneration,

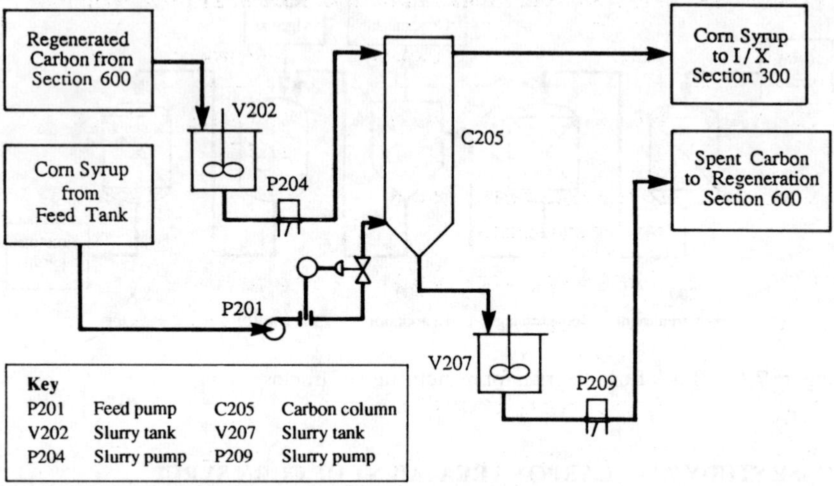

Figure 7.2 Fructose syrup carbon treatment section 200.

some carbon is lost by combustion and must be replaced from time to time by make-up carbon. How is this to be added to the section?

Carbon is delivered as pellets in multi-wall paper sacks which could be charged either directly to the column C205 or into the slurry tank V202. The former arrangement requires an elevated charge hopper and star feeder with means of lifting bags to a charging floor at a suitable height, for example by a goods hoist or fork lift truck. Weather protection and, if pellets are dusty, a dust extraction unit, may be required. There must be provision for receiving deliveries of bags from suppliers, for their storage, and for charging their contents into the slurry tank. The preliminary flowsheet Figure 7.2 shows no facilities for charging new carbon so that, in this respect, communication to plant constructors and operators is incomplete.

VESSEL CAPACITIES

These are summarised in Table 7.1. Vessels are sized for 50 kt/yr plant and are to conform to BS 5276 Part 4, eg calculated diameter for the carbon bed in item C205 is 3.12 m, adjusted to 3 m. Axial length to conform to BS is either 3 times diameter, giving a volume of 63.6 m^3, or adjusted from capacities given by

TABLE 7.1
Fructose syrup carbon treatment section 200: summary plant item list

Item	Title	Capacity	Dimensions	Material
P201A,B	Feed pumps	16.6 te/h	10.4 m head, 1.5 HP	316 SS
V202	Carbon slurry tank, covered and agitated	4.4 m^3		304L
P204	Carbon slurry pump	2 te/h	20 m head	CS
C205	Carbon treatment column	72 m^3	3 m × 12 m	304L
V207	Spent carbon tank	4.4 m^3		CS
P209	Carbon slurry pump	2 te/h	20 m head	CS

process design calculations, 4 × diameter, giving a capacity of 84.8 m^3. 72 m^3 is the calculated carbon volume, but free space is required above and below the bed.

The conical base reduces the required cylindrical volume so that the height finally specified is 4 × diameter, or 12 m.

VESSEL CONSTRUCTION

To avoid possible sites for bacterial growth and contamination, plant items are internally polished.

CASE STUDY 7.2 — REGENERATION OF ION EXCHANGER

OPERATIONS AND DESIGN OBJECTIVES

After decolourisation with active carbon, corn syrup from block 200 is passed through the ion exchange section, block 300, to reduce the concentration of calcium and magnesium ions. The I/X resin requires regeneration with salt solution every few days and the flowsheet should show provision for all operations associated with this.

The following questions arise from consideration of operational problems of resin regeneration:

(1) Can continuous operation be maintained despite interruptions for regeneration? If so, how?

(2) What arrangements are required for preparing and supplying the salt solution?
(3) When regeneration starts, should salt solution be admitted immediately?
(4) What kind of water supply is required for this section?
(5) What happens to the 'spent' salt solution?
(6) How is salt solution moved through the I/X column?
(7) When the resin has been reactivated, should corn syrup be re-admitted to the column immediately?

The flowsheet and process description should communicate to the operators how the designer proposes to resolve these questions. Feasible solutions might be:

(1) Corn syrup can be treated continuously in the ion exchange section by having two exchange columns operating alternately. While one is on-stream, the second is being regenerated.
(2) Salt solution is made in batches, each sufficient for one regeneration, by charging bags of salt into an agitated, heated solution tank. Means of raising bags to a charging floor are required, eg a goods hoist or fork lift truck.
(3) Before starting regeneration and admitting salt solution to the I/X column, as much as possible of the remaining syrup is recovered for economic reasons and to reduce effluent volume. Syrup is therefore displaced from the resin bed by water and the resulting dilute syrup is collected in a tank for recovery.
(4) The main purpose of ion exchange is to remove Ca and Mg ions. Yet towns' water, especially if hard, contains Ca and (frequently) Mg salts. Water likely to mix with syrup is therefore treated to remove these; demineralised water is specified. The salt specification should also be checked. While such matters are largely a chemist's concerns, the designer must specify appropriate equipment and be aware of the effect of all substances handled on materials of construction.
(5) After passing through the bed, spent salt solution flows to effluent drain.
(6) Salt solution is pumped downwards through the ion exchange bed leaving the bottom outlet to a stand-pipe with siphon break, which ensures that resin is covered with liquid at all times.
(7) When regeneration is complete, residual salt solution is displaced by demineralised water to avoid contaminating the syrup when normal flow resumes. These washings are also passed to effluent. When free from salt, water is displaced using dilute syrup recovered in Operation 4. Excess dilute syrup is passed to the evaporation section.

FLOWSHEET DEVELOPMENT AND PROCESS DESCRIPTION

How can these operations be developed into a flowsheet most conveniently?

One way is to sketch quickly the arrangement of plant items required for each operation and then to combine the sketches into a single flowsheet, making any modifications necessary to secure an economical operable system.

EXERCISE

7.1 Consider the operations and prepare a flowsheet to perform them. Item sizes are not required.

REFERENCES IN CHAPTER 7

7.1 Rose, J.C., Wells, G.L. and Yeats, B., 1978, *A guide to project procedure*, IChemE, Rugby, Chapter 18.

7.2 Wells, G.L., Seagrave, C.J. and Whiteway, R.M.C., 1976, *Flowsheeting for safety*, IChemE, Rugby, Figure 1.1.

Technical information about the 'Sweetzyme' glucose isomerase (enzyme)-based process discussed here was obtained from Novo Industri A/S, Bagsvaerd, Denmark. This, and principles outlined above, are incorporated in *High fructose corn syrup*, an undergraduate Student Design Project Report 1977/78, Chemical Engineering Department, University of Edinburgh.

8. LAYOUT

Trial of various plant layout schemes not only helps selection of the optimum physical arrangement for operability and maintenance, but often leads to refinement of the flowsheet also.

The physical layout of a chemical process plant must, as far as possible, satisfy in the most economical way the requirements of construction, maintenance and convenient and safe operation. These sometimes incompatible needs are best accommodated by the familiar iterative design procedure of making a trial layout, noting its deficiencies, and reducing these by successive stages of improvement.

In each such improved arrangement of plant items, the fundamental requirements of the process flowsheet must of course be satisfied, but it is generally found that, as the layout is progressively refined, significant changes in the flowsheet appear desirable. Thus, developing a good plant layout is an important step in finalising the process flowsheet and in the preparation of subsequent line diagrams, as well as in making hazard and operability studies.

TRIAL LAYOUT BENEFITS

In a logically arranged flowsheet, the position in which plant items are shown reflects their relative locations on the plant. These positions are, however, not always practicable or convenient in plant layout. For example: furnaces must be separated from process items which contain large flammable liquid inventories; distillation columns, pumps and other items performing common functions may be conveniently installed as a single group, giving operational and maintenance advantages. But changing the positions may change the number or specification of plant items on the flowsheet. As an example, the rating of a furnace used to preheat the feed to a reactor had to be increased when re-sited at a safe distance, because of unavoidable heat loss in a long pipeline.

The trial layout is thus useful in:

- helping to ensure that all main and subsidiary items of plant equipment have been identified and adequately specified;
- refining the process flowsheet;

- improving understanding of how the completed plant will operate;
- providing a basis for drawing engineering line diagrams which can be used for hazard and operability studies;
- facilitating the selection of appropriate Lang factors for preliminary capital cost estimates.

What prior information is needed to assemble such a layout model?

DATA FOR A TRIAL LAYOUT

- the draft process flowsheet;
- the accompanying process description;
- the draft plant item list showing unique item numbers, descriptions and leading dimensions;
- notes summarising any special features of plant items, such as access requirements for charging or discharging or for maintenance;
- outline vessel sketches for main plant items, particularly if these are non-standard.

Although detailed vessel design is a specialised task, designers should make preliminary proposals to ensure that these can be developed into practical designs. Features to show on vessel sketches include manways, charge holes and pipe branches. Reference 8.1 gives good examples. Vessels must be capable of internal inspection and perhaps repair. Flanges necessary for the insertion of agitators and footstep bearings, or tray supports and other internal fittings, must be shown. The way vessels are supported must also be known in order to secure satisfactory layout. Standard texts show examples of the use of such supporting means as skirts (for towers), plinths (for pumps and tanks) and structural steelwork. The check list in Table 8.1 (see pages 72 – 73) may be useful.

TRIAL LAYOUT ASSEMBLY

A trial layout is best prepared using cut-outs to represent the plan/ elevation of main plant items. (Wallpaper and plant item silhouettes cut from coloured paper can conveniently be pressed into service for layout plans!) Add item numbers prominently. Following the materials flow sequence of the flowsheet, assemble cut-outs on backing board or paper to show plan and elevation on a conveniently large scale. Observe the general constraints indicated in Table 8.2 (see pages 74 – 75) and those specific to particular cases in References 8.2 and 8.3, allowing access for erection, operation and maintenance around, above and below items as appropriate. (A convenient 'default distance' in plan is 1 m.)

Decide what structure and buildings are required to support or shelter the plant. Suitable modules for buildings allow vertical support columns at

TABLE 8.1
Check list for outline vessel design

Operating and design conditions:	The specified temperature and pressure should allow for: • Vessel design temperature usually not less than 15°C above maximum operating temperature; • Vessel design pressure usually not less than 110% of maximum operating pressure; • Start-up and shut-down; • Process upsets and surges; • Consequences of associated equipment failure; • Possible discharges of liquid from unvented vessels, leading to partial vacuum with potential vessel collapse; • Aqueous and volatile fluids heated by internal coils, where abnormal temperature rise or vessel isolation would generate excess pressure.
Initial specifications:	• Allow adequate vapour space above operating level and a design margin above the calculated process volume. • Give horizontal or vertical configuration with economic BS length-to-diameter ratio. • Specify agitator type, if required.
Pressure relief:	• Vessels and pipeline sections should be connected to, and incapable of isolation from, pressure relief devices at all times. • Large, low pressure storage tanks generally need vacuum relief. • Heat exchanger shells may be exposed to tube failures.
Vessel attachments:	• Specify branch numbers and sizes, to include requirements for process, instrumentation and access. • Minimum branch diameter is 5 cm for strength. • Manways shall be at least 60 cm but may well need to be larger to accommodate removable internals, eg agitators and trays. Full diameter removable covers are preferred to manways for small vessels. • Inlet pipes to storages should be below liquid surface, avoiding free fall. • External and internal ladders may be necessary on large storage vessels.

TABLE 8.1 (continued)
Check list for outline vessel design

Materials of construction:	• Examine possibilities of corrosion and chemical attack in relation both to main components of process streams and to trace impurities. Remember that product discolouration or contamination through chemical interaction with containing walls may be as important as vessel corrosion. • Consider effects on structural properties of temperatures at the extremes of the operating range — low as well as high, since brittle fracture may have as devastating effects as high-temperature creep.

Perry's Handbook and special texts must be consulted for data on these materials questions.

regular intervals between 5 and 7.5 m with, say, 5 m floor intervals. The first floor may be 6 m above datum with a mezzanine at 4 m if required. Top floor clearance may be reduced to 4 m.

Establish the road and emergency access to the plant block. Crane access must be provided for large plant items, not only for initial installation but also generally for subsequent servicing or replacement. Now arrange the cut-outs to fit the structural grid, allowing space for access and stairs, pipe racks, drains, aisles, roads etc. Ground floor items may have local access. Mark out the areas designated for pipe runs.

Finally check the layout for operability and economy, and make appropriate additions and amendments to the flowsheet. A trial layout exercise on a student design project, 'Esterifying terephthalic acid', resulted in the 25 items originally shown on the preliminary flowsheet being increased to 81, after considering materials handling and operational problems. A further 32 items were added to deal with minor process streams whose disposal was not considered on the original flowsheet.

Consult Reference 8.2, Chapter 2, particularly Table 4.2, 'Typical Clearances', and further information in Part 2. The expanded version of this, Reference 8.3, is encyclopaedic and appropriate passages should be consulted. Chapter 6 of Reference 8.4 has a case study, 'Layout and warehouse requirements for batch processing'. Reference 8.5 shows excellent sketches and sections for various layouts.

TABLE 8.2
Some general constraints on plant layout

Pumps and ground floor items:	• All to be supported on plinths to protect from spillages and to facilitate replacement. • Pumps require short suction lines with adequate NPSH. • Clear access is necessary for maintenance. • Ensure that drips from leaking pump seals cannot damage nearby equipment. • Ensure that cable runs are protected as far as possible from fire hazards.
Vessels:	• Require adequate headroom for installation and maintenance, especially removal of covers, agitators and drives. An empty lifting well may also be required within a structure.
Heat exchangers:	• Should where possible be mounted horizontally at or near ground level, with adequate access to remove the tube bundle. • Piping connections are normally concentrated at one end, requiring an even number of tube passes. • Vertically-mounted heat exchangers require lifting clearance for the removal of tube bundles.
Lifting gear:	• Packages of chemicals, etc which are to be lifted above 4 m require a goods hoist with room for head gear. Such a hoist may also be required for maintenance.
Aisles:	• Are to be unencumbered, 2.5 m wide. • Upper floors over 7.5 m long need a staircase at each end, each occupying half a bay with broken flights between floors.
Furnaces:	• To be separated from flammable materials.
Bulk storages and tank farms:	• Are generally outside plant structures and, for flammable or toxic materials, widely separated from plant. • Segregate dissimilar hazards. • Arrange piping and access along one side, leaving the opposite side for fire-fighting or emergency access. • The volume of bunds should be not less than 110% of the retained tank capacity.

TABLE 8.2 (continued)
Some general constraints on plant layout

Day tanks:	• Day tanks for raw materials and product testing may be sited adjacent to manufacturing areas. • Avoid simultaneous 'make and take' from the same tank by using at least two tanks for each duty.
Buildings:	Are required to house: • Control room; • Toilets; • Office; • Stores; • Local maintenance facilities.

CASE STUDY 8 — FRUCTOSE SYRUP

In this case study we consider some layout problems which are likely to arise from flowsheets for a fructose syrup plant derived in Case Studies 7.1 and 7.2. Figures 7.2 (see page 66) and A7.3 (see pages 116–117) are the flowsheets previously developed for the carbon treatment and ion exchange sections of this plant. What constraints on layout are particular to these flowsheets?

Table 8.3 lists some constraints and problems arising in these process sections.

TABLE 8.3
Fructose syrup plant: some constraints on layout of sections 200 and 300

(1) Covered storage is necessary for bags of new carbon pellets and salt. These require lifting to a charging floor above items V202 and T311. These vessels are to provide P204 and P312 with sufficient NPSH, so that charging facilities at first floor level are indicated.

(2) The flowsheet shown in Figure A7.3 indicates no means by which resin can be charged to I034, when required, nor discharged when finally spent. Column bases should therefore be at least 2 m above datum with charging again at first floor level.

(continued overleaf)

TABLE 8.3 (continued)
Fructose syrup plant: some constraints on layout of sections 200 and 300

(3) The installed cost of plant items is affected by the type of structure. What structure is required for the carbon treatment section and what is the height? In Figure 7.2 the carbon column C205 is above the slurry tank V207 which is above the pump P209. The column is 12 m high so that occasional maintenance access is necessary at third floor level.

(4) Sterility must be maintained in all plant items , so good standards of weather protection are required, indicating a building as well as a structure to support plant items. Floors should be tiled and well drained.

(5) The regeneration furnace is gas fired. Although not unusually hazardous, it is located downwind, isolated from syrup processing units so that carbon treatment columns cannot be close to the furnace. Yet carbon pellets are conveyed between items by fluidised transfer systems, requiring that lines be short and bends gradual. Transfer pipes should therefore have adequate slope, say 1 in 10, with items at adequate height to allow this.

EXERCISE

8.1 Make a trial layout for these sections of the plant. List any modifications likely to require further consideration to make the flowsheet acceptable.

REFERENCES IN CHAPTER 8

8.1 Rose, J.C., Wells, G.L. and Yeats, B.H., 1978, *A guide to project procedure*, IChemE, Rugby.

8.2 Mecklenburgh, J.C., 1973, *Plant layout*, Leonard Hill/IChemE. Note particularly the typical clearances given in Table 4.2, Chapter 2, and in Part 2.

8.3 Mecklenburgh, J.C., 1985, *Process plant layout*, Leonard Hill/IChemE.

8.4 Baasel, W.D., 1976, *Preliminary chemical engineering plant design*, Elsevier, New York. Chapter 6 has a case study on layout and warehouse requirements for a batch process.

8.5 Plant layout, Reprint 004, *Chemical Engineering*, 1977/78.

8.6 Sinnott, R.K., 1983, *An introduction to chemical engineering design*, Pergamon Press, Oxford. Chapter 14 gives a brief overview on layout.

8.7 Backhurst, J.R. and Harker, J.H., 1973, *Process plant design*, Heinemann, London. Note Chapter 11, 'The complete plant'.

9. LINE DIAGRAMS

Engineering line diagrams are the most elaborate and detailed plant drawings for which the chemical engineer is normally responsible.

NATURE AND FUNCTION

Engineering line diagrams (ELDs) are more complete representations of chemical plant than are process flowsheets. As can be seen from the specimen portion of such a diagram reproduced in Reference 9.2, they show all the pipelines, fittings and control equipment which can be seen on a completed chemical plant. These details will, of course, themselves require separate drawings, as will the layout and structural particulars of process and services piping and instrument and electrical wiring. Detailed drawings of the latter kind are not, however, primarily the responsibility of the process designer, but of mechanical and electrical design specialists. The ELD represents the final stage of design elaboration for which the chemical engineering designer is normally directly responsible.

Line diagrams are essential for hazard and operability studies, for training operators and, of course, in the later stages of design work, for improving the accuracy of earlier capital cost estimates.

To construct an ELD, we need a process flowsheet, a process description, a plant item list with summarised vessel specifications, and a proposed layout to show distances and relative positions of plant items.

A SYSTEMATIC PROCEDURE

Professional designers may use a modular approach to ELD construction. They know from experience what lines and arrangements are required for standard equipment items (eg a distillation column and ancillaries); and by adding to each major equipment item on the flowsheet such associated groups of ancillaries they are able to build up the whole diagram.

The principles of line diagram construction are, however, better illustrated by a painstaking systematic approach through the following steps:

(1) Examine the flowsheet to define, for each item, the design objective, the physical and chemical changes taking place and the operating constraints.
(2) Check that main process variables, including composition, are adequately controlled and that manipulated variables are known.
(3) Consider the susceptibility of each item to fouling, corrosion or failure and the effect on the system of its possible unreliability. Decide what installed spares are necessary and what can be covered by workshop spares and maintenance policy.
(4) Consider what provision should be made for inter-stage storage capacity, to ensure smooth continuing operation in the face of minor malfunctions.
(5) Consider all operations which have to be performed and draw the necessary pipelines on the diagram. This should be spaciously drawn using approximately scaled and positioned items.
(6) Insert all valves, meters, controllers, etc required for mass flow and control.
(7) Make similar additions for temperature measurement and control.
(8) Find what determines the pressure in each plant item and how this might vary in normal or abnormal operation. Consider how valve closures might isolate plant sections containing fluids which could cause overpressure. Make additions for pressure measurement, control and relief from possible overpressure. Some reconsideration of earlier versions of the diagram may be necessary at this stage.
(9) The properties of the materials handled at each stage may indicate the need for additions; eg hydrocarbon/air mixtures can be flammable and explosive and so need safe pressure relief and venting. Nitrogen purging and blanketing may be required, eg in storage tanks.
(10) Design calculations usually concentrate on operations at flowsheet rate with feeds of specified composition. Lower rates or different feedstock are often used. Examine the ELD to see whether such modes of operation require special provisions to be made.
(11) Additions may be necessary for a controlled shut-down of the plant, ie, when enough product has been made, or the plant has to be emptied and cleaned for maintenance. Start up may require further additions.
(12) Plant item failure and external events such as fire or mal-operation may require emergency shut-down and the safe disposal of the plant inventory. Additions should be made for any extra vessels and pipelines which this would require.
(13) Iterate and check that the ELD is still satisfactory for all operating modes. Shade in those valves normally closed. Indicate whether, in the event of air

failure, control valves should move to the open or closed position or should stay in their last position.
(14) Confirm the design and the diagram by completing an operability study as described below.

OPERABILITY STUDIES

Operability studies enable finished line diagrams to be checked for completeness and acceptability, to ensure plant safety and to train operators. The systematic procedure is well documented in Reference 9.5. A summary of the procedure is:
(1) Set up a small team drawn from representatives of the plant operations management, with engineers from the design, technical, project, instrument and control and maintenance functions.
(2) Examine the line diagram with its associated process description line by line, section by section, for the possibility and implications of deviations from the designer's intentions, using the following guide words to classify such possible deviations:
NONE: eg no flow;
MORE OF: eg flow, temperature, pressure etc;
LESS OF:
PART OF: eg some component concentration lower than intended;
MORE THAN: eg impurities (formed by decomposition or otherwise) or extra phases present;
OTHER: ie abnormal conditions of operation, such as start-up or shut-down, catalyst failure, maintenance, etc.
(3) List the consequences of each deviation and evaluate whether these: (a) are acceptable, so that no further action is required; (b) are not acceptable, so that immediate change is necessary; (c) require further study to reach a conclusion.
(4) List actions required.

Full details and examples are found in Reference 9.5.

The development of line diagrams and their use in operability studies is illustrated in the following case study in which a hydrocarbon mixture is prepared for reaction with feed/product heat interchange.

CASE STUDY 9 — ALKENE DIMERISATION: FEED PREPARATION

ALKENE DIMERISATION: PROCESS INFORMATION

The objective of alkene dimerisation is to produce high octane hydrocarbons which enhance the value of gasoline. Anhydrous aluminium chloride, a Friedel-

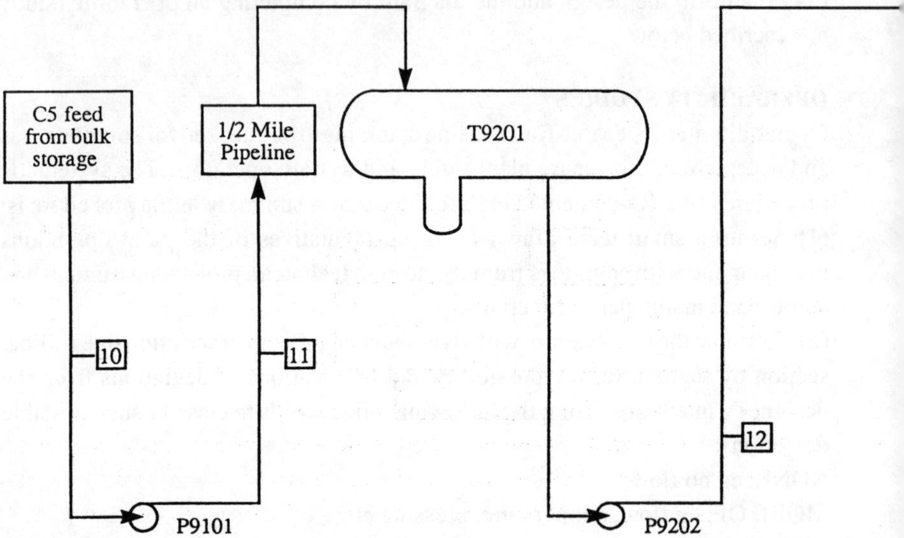

Figure 9.1 Alkene dimer: abbreviated flowsheet.

Crafts catalyst, may be used in a fixed bed reactor under carefully controlled conditions, which exclude traces of water. General process information is disclosed in Reference 9.3.

The following is a brief process description of Figure 9.1, an abbreviated process flowsheet for the feed section of such a plant.

An alkene/alkane fraction, which may occasionally contain small amounts of suspended water, is pumped continuously by P9101 from bulk storage through a pipeline half a mile long into feed tank T9201. This acts as a buffer and settling tank where any traces of water are settled out and run off at intervals. P9202 then pumps dry hydrocarbon to a feed/product heat interchanger H9203, a preheater H9204 and thence to the reactor R9205, where residence time is held within close limits to secure adequate conversion while avoiding polymer formation and consequent fouling of heat transfer surfaces.

Feed composition is variable and includes pent-1-ene, pent-2-ene with pentane and some C_4 and C_6 components. Physical properties of these are found in standard handbooks. Table 9.1 gives stream data for Figure 9.1.

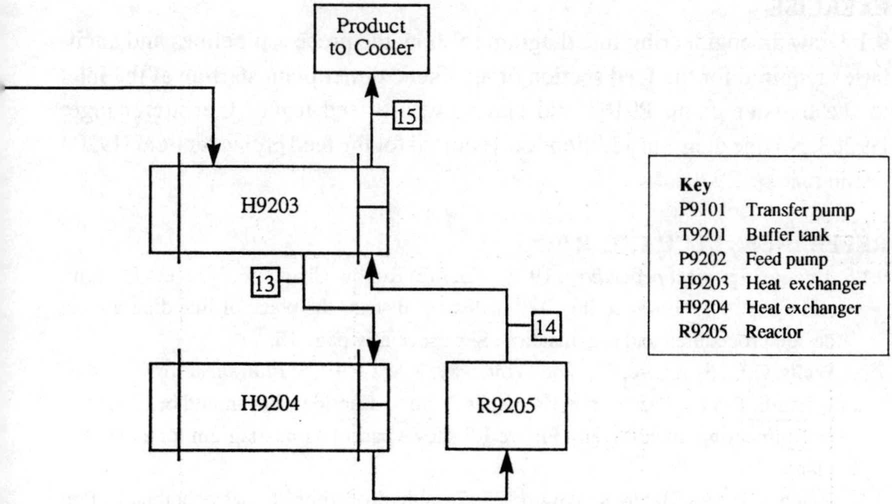

TABLE 9.1
Stream data for alkene dimer flowsheet

	Stream	°C	bar
11	P9101 transfer pump outlet	20	2.4
	T9201 buffer/settling tank contents	20	1.05
12	Feed stream at H9203 inlet	20	20.7
13	Feed stream at H9203 outlet	160	20.0
14	Product leaving reactor R9205	205	17.9
15	Product leaving H9203	60	16.6

EXERCISE

9.1 Draw an engineering line diagram to show the process pipelines and ancillaries required for the feed section of an alkene dimer plant starting at the inlet to the transfer pump P9101 and ending where feed leaves the interchanger H9203. No line diagram additions are required for the feed preheater item H9204 or the reactor R9205.

REFERENCES IN CHAPTER 9

9.1 *A guide to project procedure*, 1978, IChemE, Rugby. Chapter 6, 'Process Design', and Chapter 7, 'Engineering Design: Basic' discuss the place of line diagrams in design procedures and organisation. See especially page 15.

9.2 Wells, G.L., Seagrave, C.J. and Whiteway, R.M.C., 1976, *Flowsheeting for safety*, IChemE, Rugby. Paragraphs 1.4, pages 2 and 4, describe what should be shown on an engineering line diagram. Figure 1.2 shows part of a line diagram for a benzene plant.

9.3 Hatch, L.F. and Matar, S., Aug. 1977, From hydrocarbons to petrochemicals, Part 4, *Hyd Proc,* 56: 155.

9.4 Scott, R., February 1982, *Drawing an engineering line diagram — a plant design workshop*, University of Edinburgh Chemical Engineering Department Course Manual.

9.5 Lawley, H.G., 1974, Loss prevention: operability studies and hazard analysis, *Chem Eng Prog*, 70 (4): 45–56.

9.6 Dead head hazards of centrifugal pumps, IChemE *Loss Prevention Bulletin* 029, 139.

10. OPTIMAL DESIGN

THE CONCEPT OF PROCESS OPTIMISATION

The prospective profitability of a process scheme is measured by the difference between what is realised by the sale of the product and the costs of production (including, of course, capital and depreciation charges). Usually, some at least of these costs are sensitive to changes in operating conditions; the profitability will then generally be greatest for some particular set of operating variables. If the full competitive commercial potential of the process is to be fairly assessed, the optimum variables should be identified and used in cost estimation. Under these conditions the process design is optimised with respect to overall profitability.

❒ Q. *Is there more than one basis on which a process design may be optimised?*

The set of operating variables which gives minimum total cost does not in general minimise each of the individual costs making up the total. In particular circumstances, some sub-set of costs (or even some individual cost item, such as that of toxic waste disposal) may assume special, or controlling, importance. The optimal operating conditions are then those which minimise this sub-set of costs. Thus, the sense in which a design is to be optimised requires careful consideration of the operating circumstances; and the designer must state explicitly what are the operating variables or combinations of such variables — overall profitability, capital cost, foreign exchange cost, energy consumption, freedom from hazard, to name but a few — with respect to which he has optimised his design.

PROCEDURE

Where there is little previous experience to draw on, the designer of a new process will adopt a 'base case' for his first design scheme, having the firmest attainable technical background and following closely those operating conditions for which calculation originally showed the process to be viable. He will then examine the possibilities for improving on this. Part of the skill of the designer is the ability to spot changes in the original conception which are likely to yield improvements and cost reductions; but these intuitive leads must be followed and supported by detailed calculations from the earliest stages, because changes in a complicated design scheme which lead to cost reductions of one kind lead in general to consequential cost increases elsewhere. There are usually no short-cut methods for finding the most advantageous trade-off; this can only be done by a careful analysis of the effect of each proposed design change on individual costs.

❐ Q. *How can the designer make preliminary cost estimates so that he can begin such an analysis?*

COSTS

The principal operating costs for a proposed process can be estimated as soon as the mass and energy balances are calculated. Plant costs, and hence capital expenditure, cannot, however, be determined until at least a preliminary design has been arrived at and the capacities of individual plant items are known. A preliminary estimate of project profitability can then be made by subtracting the annual costs attributable to operation and capital investment and depreciation from expected annual revenue from product sales.

The effect of changes in the operating variables on this preliminary estimate of yearly profit should then be explored, so that a rough estimate of optimal profitability can be made. This can then be compared with the sponsor's financial criteria for proceeding with such schemes. The preliminary estimate of optimal profitability can also be compared with that of alternative courses of investment and action, such as the manufacture of other products, buying in product for resale, manufacture by different processes, building the manufacturing plant of different capacity, or in another location, from that originally envisaged.

This preliminary procedure, taking account only of the most salient items of cost and revenue, can be relatively quickly accomplished and may yield immediate and important information about the viability or otherwise of the process. If the indications from it are that the proposed process is likely to be commercially attractive, essentially the same procedure must next be undertaken more exhaustively, with greater attention to consequential costs and secondary constraints on plant construction and operation.

The symposium of Reference 10.1 gives guidance on these more detailed matters, and is concerned with optimising various aspects of project selection, process design and plant operation. Techniques for optimising energy usage, including the application of process energy integration, are given in Reference 10.2. Here, however, we return to more basic questions about how to initiate an optimisation procedure.

Case Study 10, cyclohexanol manufacture, illustrates how capital cost and operating cost may be optimised in a recycle process where conversion and selectivity are the main process variables. This cyclohexane oxidation process can be criticised for the hazard created by its high inventory of flammable liquid and unusually low conversion consequent upon an inherently slow stage in a complex series of reactions. For these reasons the process has become obsolete; but it is a good example for our purpose, because data needed for optimisation calculations are in this case available in the literature.

CASE STUDY 10 — CYCLOHEXANOL MANUFACTURE

In Case Study 5 we constructed a process for isolating cyclohexanol/cyclohexanone from the crude mixture resulting from cyclohexane oxidation. Figure 5.1 on pages 54 and 55 is the resulting schematic flowsheet.

Experimental work shows that at 7.8% conversion of cyclohexane, selectivity to desired product is 65.8 mol percent and a technically firm 'base case' design is based on these conditions. Table 10.1 is a summary plant item list with estimated costs (see pages 86 – 87).

❒ Q. *What would be the consequences of changes in the chosen conversion and the attendant changes in selectivity? What information would you need to answer this question?*

CONVERSION AND SELECTIVITY

The low conversion of the base case design (7.8%) results in very high cyclohexane recycle and in high capital costs for large equipment in blocks 300 and 700, the reaction and distillation sections. Alternative designs for higher conversions are therefore worth investigation. For these, however, the selectivity will in general be less; for, as shown in Figure 10.1 on page 87, selectivity falls as conversion rises. Fundamentally, this is because the desired product then undergoes further oxidation. What effect has this on stage capacities and costs?

To answer this, we must examine the schematic flowsheet and tabulate the effect on each section of designing for higher conversion. Designing for higher conversion shows that significant changes occur in sections:

TABLE 10.1
Cyclohexanol plant: summary plant item list to produce 6.25 te/h at 7.6% conversion

Item No.	Description/Duty/Special features	Cost, 000£, base year
101 – A,B,C	Cyclohexane storages.	*
102	Catalyst handling.	*
201	Air filter.	24
202	Air compressor and drive. 21 te/hr to 10 bar in 2 stages. 1.3 MW.	276
301 – A,B	Feed pumps.	20
303 – 7	Reactors. 1.2 hr total hold-up. 3.5 m dia × 6.5 m. 10 bar. SS. Agitated. Air spargers. 0.5 m dia overflow.	350
401 – 5	Off-gas condensers. 15 MW total. Shell and tube. 80 m^2 ea. SS/SS.	150
407 – 11	Reflux separators.	*
501	Absorber. 1 m × 10 m, 2.5 cm packing. SS.	35
502	Oil circulation tank.	*
503	Oil circulation pump.	*
504 –	Stripper and auxiliaries.	*
601, 3	In-line mixers.	*
602, 4, 5	Liquid/liquid separators. 10 mins hold-up ea. 2 m × 7 m. SS.	65
701	Distillation column and plates. 2.7 m × 12 m. 10 sieve trays. SS.	125
702	Condenser/cooler. 10 MW. Shell and tube. 400 m^2. SS/SS.	80
703	Reboiler. 500 m^2. U-tube. SS.	100

TABLE 10.1 (continued)
Cyclohexanol plant: summary plant item list to produce 6.25 te/h at 7.6% conversion

Item no.	Description/Duty/Special features	Cost, 000£, base year
705 – A,B	Reflux pumps.	*
706	Reflux drum.	*
707	Cyclohexane reflux tank. 1 hr hold-up. Fixed cone roof. 200 m^3.	120
708 – A,B,C,D	Product tanks.	*
800	Central waste disposal facility charged at operating cost plus return.	*
Total variable plant item cost.		1345

* Cost varies little with conversion.

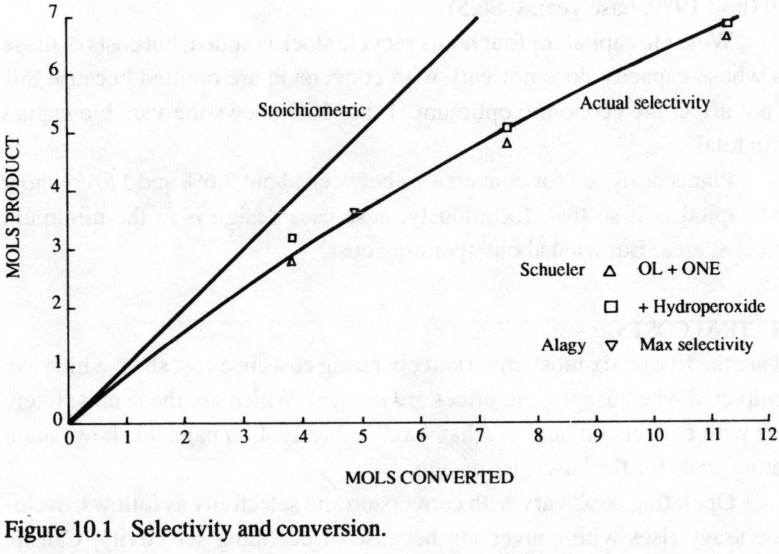

Figure 10.1 Selectivity and conversion.

• 200 (air filter, compressor and drive) because selectivity falls and more air is used in oxidising cyclohexane to acid than to cyclohexanol;
• 300, 400 and 500 (reactors, condensers and absorber) because more disengaging area is needed for the greater nitrogen volume and the acid production is more exothermic than is cyclohexanol production;
• Less capacity is required at stages 600 (separators) and 700 (excess cyclohexane distillation).

Which should be examined first, operating costs or capital costs?

CAPITAL COSTS

Capital costs are examined first, because in making new plant proposals to management, it is in the first place capital expenditure above the minimum that must be justified. Capital costs depend on plant item costs and these vary with the capacity of each item according to the standard expression:

Cost α (Capacity)^n

Table 10.2 shows how plant item costs vary with capacity, giving values of the exponent *n* for different cases.

How are total fixed capital costs derived from plant item costs? Plant item costs are converted to installed capital costs by multiplying by:
• overall Lang factors, totalling 4.4 for this type of plant;
• a cost inflator for chemical plant construction (estimated, for example in 1987, at 119 (mid 1979, base year, = 86.5)).

Working capital for four hours recycle stock is added; but costs of those items whose capacity does not vary with conversion are omitted because this does not affect the economic optimum. Table 10.3 shows the variable capital costs in total.

Plants designed for conversions between about 7.6% and 11.4% show lowest capital cost so that, fortuitously, base case design is in the minimum capital cost area. But what about operating cost?

OPERATING COSTS

What are the five or six most important operating costs in a cost sheet which can be completed when usages and prices are known? Which are those most likely to vary with conversion, and in what way? Table 10.4 on page 90 shows main operating costs for the base case design.

Operating costs vary with conversion and selectivity as follows: cyclohexane usage rises with conversion because of declining selectivity. Caustic

TABLE 10.2
Plant item cost variations

Item No.	Description	Capacity determinant		Exponent
		Stream	Material	
201	Air filter	4	Air	0.67
202	Compressor and drive	4	Air	0.8
301	Feed pumps	3	Feed	0.67
301 – 7	Reactors	4	Air	0.9
401 – 6	Condensers	4	Heat	0.67
501	Absorber	4	Air	0.67
602 – 5	Separators	9 + 2 + 1	Total flow	0.67
701	Distillation column and plates	2	Recycle	0.67
703	Reboiler	2	Recycle	0.67
707	Recycle tank	2	Recycle	1.0

TABLE 10.3
Costs which vary with conversion

Conversion, %	Capital cost, 000£	Operating cost, £/te product
11.4	10,040	760.2
10.0	10,045	738.0
7.6	10,145	697.4
6.0	10,397	667.7
5.0	10,763	661.6
3.8	11,604	657.8
3.0	12,688	673.3

usage and waste disposal charges rise similarly, as does the smaller cost for electricity used for air compression. The principal cost which falls at higher conversion is that for steam for cyclohexane distillation.

Table 10.3 showed operating costs at various conversions alongside capital costs. Operating cost is lowest at 3.5% conversion but capital is lowest above 7.6%. How do we decide what conversion to use for design?

OPTIMAL DESIGN

Capital expenditure which is above the minimum technically acceptable must be rewarded by savings in operating cost. The quickest way of evaluating these is by incremental pay-back, a method described in Reference 10.3. Two or three years is currently an acceptable maximum pay-back time for optional extra capital. The design basis which meets this criterion has a conversion of around 5%. Subject to confirmation of product selectivity, this is the recommended basis.

TABLE 10.4
Principal operating costs for base case design

Raw materials		**Usage, t/t product**	**Price, £/te**	**£/te, product**
Cyclohexane		1.29	400	516
Caustic as 100% NaOH		0.42	150	63
Material cost				579
Variable expenses	**Units**	**Usage**	**£/unit**	**£/te**
Steam	te	0.2 t/t b/u	13	61
Electricity	MWh	1.3	35	7
Waste disposal	te act. wt	4.99	10	50
Operating cost				697

REFERENCES IN CHAPTER 10

10.1 Scott, R., 1987, 'Optimal cost design of a cyclohexane plant', in *Process Optimisation*, Symposium Series No. 100, IChemE, Rugby.

10.2 Sinnott, R.K., 1983, *An introduction to chemical engineering design*, Pergamon Press.

10.3 Allen, D.H., 1991, *A guide to the economic evaluation of projects*, IChemE, Rugby.

11. DESIGN SUCCESS

DESIGN AND CAPITAL PROJECTS

Manufacturing organisations renew and revitalise themselves through capital projects. Plant and equipment wear out, processes become obsolete and, most significantly, market demands change; for these and other reasons manufacturers must constantly devise and undertake new schemes of capital investment simply to remain in business. Ideas for such projects for investment will originate from all sections of a thriving company and, particularly in times of expansion, a successful organisation will inevitably generate more ideas than it has the resources to implement.

Evaluation and selection procedures must therefore be developed to manage the flow of proposals and the implementation of new schemes. Only a limited number of proposed designs can be successful candidates for full-scale development; how can the most promising be identified?

EVALUATING THE MERITS OF A PROPOSAL

Chemical companies, it is often said, are in business not essentially to make chemicals, but to make money. The ultimate measure of success for any manufacturing operation is the extent to which it is financially rewarding to the investors (normally, the company shareholders) when all outgoings and gains are taken into account. Of two alternative proposals, it would therefore be expected that the more profitable financially would be favoured for development.

No such simple principle, however, will alone suffice in evaluating a new project. For although profitability is essential for success, other qualities listed below, equally vital for a process that is to be successfully developed for large-scale production, are not immediately or easily expressed in financial terms, and different schemes will have these qualities to different degrees. In view of this, how can one judge systematically the relative merits of alternative schemes for full-scale development?

❒ Q. *What in your view are the most significant criteria determining the ultimate success of a production process?*

CRITERIA FOR SUCCESS

Companies have different views about what criteria to use to judge the likely success of a design, and the relative importance of these will differ from project to project. In general, the author's list of six such critical process characteristics would be:

A. Safety — ie assured plant integrity, low risk to workers and the public.

B. Product quality — consistent ability to meet required product specification.

C. Reliability — low equipment failure rate, low rate of catalyst activity change etc, leading to high on-stream time and predictable, modest, maintenance requirements.

D. Operability — ability to tolerate normal variations of raw material quality, catalyst activity and operating conditions without compromising satisfactory operation, controllability and ease of start-up.

E. Predictable usages — ie product yield, consumption of materials and energy all at design levels for design throughput, with high variable/fixed ratio of energy use at other throughputs.

F. Profitability — as assessed by methods reviewed in Table 11.1 (see page 94).

Many other factors, mostly unquantifiable, also affect preference in choice of process schemes. Notable examples are:

G. Degree and nature of environmental impact — eg by chemical or thermal pollution or noise emission.

H. Sensitivity to changing commercial factors — eg market price fluctuations of feedstocks, products, energy, catalysts, etc.

I. Relation to existing operations — ie possibilities of integration with the company's other plant or products.

J. Patent or licensing considerations.

K. Company's technical experience of related processes, design bases and assumptions.

RANKING THE CRITERIA

When such criteria relevant to the choice of a candidate process scheme have been identified, they must be ranked in order of importance as an aid to assigning an overall 'figure of merit' to any particular scheme having the listed qualities in varying degrees.

A well-known technique for ranking many qualities relative to one another is the 'forced pairs' method of comparison. The qualities considered, A – Z, are written as the headings of the columns and rows of a square grid (see

TABLE 11.1
Methods of project profitability evaluation

Return on Investment (ROI)	Simple calculation of profit after depreciation on historical fixed plus working capital. (But for which year?) Useful for comparing processes. The essential deficiency of this method is that it fails to recognise the declining value of money.
Return on Capital Employed (ROCE)	Similar to ROI but on depreciated fixed plus working capital. Mainly for accounting use.
Pay-back	Shows how soon initial expenditure is recovered. May be discounted, ie corrected for the diminishing present worth of increasingly distant returns. Use for small projects and in times of cash shortage, when the pay-back time is required to be very short.
Net Present Value (NPV)	Represents project profitability as a single cash sum at present-day value after charging the assumed cost of raising extra capital. Values for different projects are additive, so NPV is useful for project portfolio planning.
Discounted Cash Flow Rate of Return (DCFRR)	Shows profitability as a single annual rate of return even for projects whose profitability varies with time, eg new ventures. Calculated sensitivity to changes in project assumptions can include forecast inflation. For almost universal application, but note that natural resource projects such as oil fields and nuclear power stations yield unusually low DCFRR values, especially when costs of terminating the project at the end of its useful life are appreciable.

Table 11.2). Proceeding now along the 'A' row, if B is ranked more important than A, insert a zero in the 'B' column of the 'A' row; if A is ranked more important, insert 1 in that space. Using the same scoring convention for each column in turn of the 'A' row, inserting 1 at each column for which A is deemed of higher rank than the quality at the head of that column, we obtain a total score for the 'A' row by summing the scores for that row. If this procedure is repeated

TABLE 11.2
Criteria ranking for successful design

Criterion	A	B	C	D	E	F	Score
A	–						
B		–					
C			–				
D				–			
E					–		
F						–	

for each row in turn, the magnitudes of the total scores for each row are in the order of importance of the corresponding qualities.

Thus considering, for example, the qualities listed above; if A, Safety, is deemed more important than any of the others it will be ranked by this method at the head of the list; if B, Product quality, is judged more important than anything else but Safety, B will appear second in this ranking order, and so on.

RANKING CANDIDATE PROJECTS

It then becomes possible in principle to use such a ranking order of desirable process attributes to rate the desirability of proceeding with, or developing, alternative process schemes on the basis of their probable full-scale commercial viability. Each such scheme will have attributes A – Z to different degrees; a 'figure of merit' can be assigned to each candidate scheme by combining the extent to which it displays each such quality with the score representing the importance of that quality obtained as above, and forming a total of such weighted scores.

It is doubtful whether such operations could ever in practice be performed literally and quantitatively, but the underlying principle is nevertheless useful as a guide in choosing process alternatives.

Thus, let us suppose for the sake of illustration that our list above does indeed place the criteria A to L in the agreed order of diminishing importance, and that a process scheme is proposed for which criterion I is particularly well satisfied, in that an existing by-product of the company's operations could be the feedstock for the proposed process. Because such a favourable relation to existing operations does not (on present assumptions) rank very high as an important process attribute, this systematic consideration shows that the proposal is not likely to carry much weight on this ground alone, attractive as the feature might seem when considered in isolation. In particular, if the proposed process is such that, say, consistent product quality cannot be readily assumed (perhaps because some difficultly separable impurity is present in the proposed feedstock), so that it is deficient in respect of criterion B, it is most unlikely to prove a success in competition with some other scheme ensuring better product quality but using a bought-in feedstock.

CONCLUSION

We return to a fundamental proposition: design is about choice; the wiser and better informed the choice, the more likely is the design to be successful — successful initially and locally with Company management in competing for limited resources for implementation and development, and successful ultimately and globally as the basis of a safe, profitable and smoothly running manufacturing operation.

Wise choices among alternative plant design schemes can only be made through a thorough understanding of how a plant as conceived would in fact operate. Particular attention is given to this matter in Chapter 6; it calls for clear imagination and analytical skill supported by a sound understanding of basic principles. These qualities are as central to the successful practice of chemical process design as is the more widely discussed capacity for innovation, or creativity. All new designs must begin with innovation, but only a small proportion of innovative thoughts lead eventually to successful designs.

It is hoped that this little book will have helped, among other things, to explain how design differs from invention. Whether or not the ability to invent can be taught, it is certain that the other, equally important, skills of the designer can be developed by teaching. It is the aim of books like this to facilitate that process.

APPENDIX I — CALCULATION OF THE EQUILIBRIUM FRACTIONAL CONVERSION IN THE REACTION OF METHANOL WITH ISOBUTENE AND ESTIMATION OF ADIABATIC TEMPERATURE RISE

PRINCIPLES

The reaction of interest is:

$$CH_3OH(l) + C_4H_8(l) \rightarrow (CH_3)_3COCH_3(l) \quad (1)$$

The symbol (l) after each species in the equation indicates that all are in the liquid phase. Representing the reactant molecules as M (methanol) and B (isobutene), and the molecule of product as MTBE, the equilibrium conversion depends on the value of the equilibrium constant K_a under the reaction conditions, defined for this reaction (involving only one molecule of each species) as:

$$K_a \equiv \frac{a_{MTBE}}{a_M \, a_B} \quad (2)$$

where a represents the activities of the substances indicated by the subscripts. These would be the same as the mol fractions of the components if the mixture were ideal; so in that case, knowledge of K_a would give directly the ratio of product to unreacted molecules at equilibrium.

K_a can be found at the absolute temperature T from the equation:

$$\ln K_a = -\Delta G^\circ_R / RT \quad (3)$$

where ΔG°_R is the standard molar free energy change for the reaction at temperature T, given for the reaction M + B = MTBE by the difference between the free energies of formation, ΔG°_f, of MTBE and M + B at temperature T; ie

$$\Delta G^\circ_R = \Delta G^\circ_{fMTBE} - (\Delta G^\circ_{fM} + \Delta G^\circ_{fB}) \quad (4)$$

Now, values of ΔG°_f are usually available in thermodynamic data tables only at 25°C. ΔG°_R may, however, be found for reaction at other temperatures from the equation:

$$\frac{d}{dT}\left(\frac{\Delta G^\circ_R}{T}\right) = -\frac{\Delta H^\circ_R}{T^2} \quad (5)$$

where ΔH°_{R} is the standard heat (or enthalpy change) of reaction, given by the difference between the heats of formation of products and reactants:

$$\Delta H^{\circ}_{R} = \Delta H^{\circ}_{fMTBE} - (\Delta H^{\circ}_{fM} + \Delta H^{\circ}_{fB}) \quad (6)$$

Further, however, it must be noted that ΔH°_{R} in its turn varies with temperature, according to the equation:

$$\frac{d(\Delta H^{\circ}_{R})}{dT} = \Delta C^{\circ}_{p} \quad (7)$$

where ΔC°_{P} , the standard heat capacity change of reaction, is the difference between the standard molar heat capacities, C°_{P}, of products and reactants. Strictly these C°_{P} values, also, are functions of temperature, but for our purposes the temperature dependence of C°_{P} can be ignored and mean C°_{P} values can be taken as constant over the restricted temperature range of interest without significant error.

A final complication arises from the fact that ΔG°_{f} and ΔH°_{f} values given in the literature for product and reactants may not always be for the states (solid, liquid or gas) of these substances appropriate to the desired reaction conditions. The relation between ΔG°_{f} values for a given substance at temperature T in its liquid and gas states is given (on the assumption that the vapour behaves ideally) by:

$$\Delta G^{\circ}_{f(l)} = \Delta G^{\circ}_{f(g)} + RT \ln p* \quad (8)$$

where $p*$ is its vapour pressure at T°. For ΔH°_{f}, values for the liquid and gas states of a given substance are related by:

$$\Delta H^{\circ}_{f(l)} = \Delta H^{\circ}_{f(g)} - \Delta H_{V} \quad (9)$$

where ΔH_{V} is the molar enthalpy change (heat) of vaporisation of the substance.

APPLICATION

The data shown on page 100 for the reaction system (1), obtained partly from Reference 1.2 and partly from proprietary sources, appear in the Data Book prepared by an MTBE process design group.

FREE ENERGIES AND HEATS OF FORMATION AND LIQUID SPECIFIC HEATS

Substance	$\Delta G^\circ_{f,298}$, kJ mol^{-1}	$\Delta H^\circ_{f,298}$, kJ mol^{-1}	$C_p^\circ{}_{(l)}$, J mol^{-1} K^{-1}
MTBE	–125.44 (g)	–292.88 (g)	190 (l)
Isobutene	58.07 (g)	–13.99 (g)	162 (l)
Methanol	–166.9 (l)	–238.94 (l)	96.5 (l)
Isobutane			133 (l)

HEATS OF VAPORISATION

Isobutene	24.96 kJ mol^{-1}
MTBE	30.47 kJ mol^{-1}

VAPOUR PRESSURE EQUATIONS

MTBE	$\ln p* = 11.0383 - 3665 / T(\mathrm{K})$, where $p*$ is in atmospheres
Isobutene	$\ln p* = 11.30619 - 3002.6 / T(\mathrm{K})$, where $p*$ is in atmospheres

From these data, we first calculate liquid phase values of ΔG°_f and ΔH°_f for MTBE and isobutene, using equations (8) and (9):

$$\Delta G^\circ{}_{f(l)} = \Delta G^\circ{}_{f(g)} + RT \ln p*$$

$$\Delta H^\circ{}_{f(l)} = \Delta H^\circ{}_{f(g)} - \Delta H_V$$

For MTBE at 298°K, $p* = \exp(11.0383 - 3665/298) = 0.2836$ atm

$$\Delta G^\circ{}_{f(l)} = -125.44 + 8.314 \times 10^{-3} \times 298 \ln 0.2836 \text{ kJ mol}^{-1} = -128.56 \text{ kJ mol}^{-1}$$

$$\Delta H^\circ{}_{f(l)} = -292.88 - 30.47 = -323.35 \text{ kJ mol}^{-1}$$

For isobutene at 298°K, $p* = 3.422$ atm

$$\Delta G^\circ{}_{f(l)} = 58.07 + 8.314 \times 10^{-3} \times 298 \ln 3.422 \text{ kJ mol}^{-1} = 61.12 \text{ kJ mol}^{-1}$$

$$\Delta H^\circ_{f(l)} = -13.99 - 24.96 = -38.95 \text{ kJ mol}^{-1}$$

Hence $\Delta G^\circ_{R,298} = -128.56 - 61.12 - (-166.9) = -22.78 \text{ kJ mol}^{-1}$

$$\Delta H^\circ_{R,298} = -323.35 - (-38.95) - (-238.94) = -45.46 \text{ kJ mol}^{-1}$$

We now evaluate ΔG°_R and ΔH°_R at reaction temperature T. For $\Delta H^\circ_{R,T}$, from equation (7) and assuming ΔC°_p constant, we have:

$$\Delta H^\circ_{R,T} - \Delta H^\circ_{R,298} = \Delta C^\circ_p (T - 298) \text{ kJ mol}^{-1}$$

Here, $\Delta C^\circ_p = 190 - 162 - 96.5 = -68.5 \text{ J mol}^{-1}\text{K}^{-1}$

Hence, $\Delta H^\circ_{R,T} = -45.46 - 0.0685\,(T - 298) = -25.044 - 0.0685T \text{ kJ mol}^{-1}$

For $\Delta G^\circ_{R,T}$, equation (5) gives:

$$\frac{\mathrm{d}}{\mathrm{d}T}\left(\frac{\Delta G^\circ_{R,T}}{T}\right) = -\frac{\Delta H^\circ_{R,T}}{T^2} = \frac{25.044}{T^2} + \frac{0.0685}{T} \text{ kJ mol}^{-1}\text{K}^{-2}$$

Integrating,

$$\frac{\Delta G^\circ_{R,T}}{T} = \frac{\Delta G^\circ_{R,298}}{298} - 25.044\,(1/T - 1/298) + 0.0685 \ln\left(\frac{T}{298}\right)$$

Substituting the value of $\Delta G^\circ_{R,298}$ obtained above,

$$\Delta G^\circ_{R,T}/T = -0.383 - 25.044/T + 0.0685 \ln T \text{ kJ mol}^{-1}\text{K}^{-1}$$

So, by equation (2),

$$\ln K_a = -\frac{\Delta G^\circ_{R,T}}{RT} = 46.00 + \frac{3012}{T} - 8.239 \ln T$$

If we start with a stoichiometric feed containing 1 mol of isobutene, 1 mol of (inert) isobutane and one mol of methanol, after X units of reaction have taken place we have $(1 - X)$ mols of methanol, $(1 - X)$ mols of isobutene, X mols of MTBE and 1 mol of isobutane. The total number of mols is now $(3 - X)$ and the component mol fractions are:

Isobutene	$(1 - X)/(3 - X)$
Methanol	$(1 - X)/(3 - X)$
MTBE	$X/(3 - X)$

The equilibrium constant K_a is defined in terms of the activities of the reactants and products in the liquid mixture, according to equation (2).

However, no data are available for the liquid phase activity of MTBE in methanol-hydrocarbon mixtures, nor for its variation with composition and temperature. We therefore approximate for K_a by substituting mol fractions for activities in equation (2) as if the system were ideal.

Then, $K \approx X_{MTBE} / X_M X_B \approx K_a$

$$K_a \approx K = \frac{X/(3-X)}{((1-X)/(3-X))^2} = \frac{X(3-X)}{(1-X)^2}$$

Expanding the quadratic and solving it for X,

$$X = \frac{3 + 2K - \sqrt{9 + 8K}}{2(1+K)}$$

In the above equation, only the required root between 0 (zero conversion) and 1 (complete conversion) is given. Finally, we can now tabulate the fractional conversion of methanol and isobutene, X, as a function of temperature, taking $K \approx K_a$:

These results, plotted more extensively in Figure 3.2 (see page 31), show the familar form for an exothermic reaction in which the equilibrium conversion decreases rapidly as temperature increases.

T, °C	T, K	K_a	X
25	298	9595	0.986
50	323	2260	0.970
60	333	1328	0.962
70	343	800	0.951
80	353	492	0.937
90	363	309	0.921
100	373	198	0.902
110	383	129	0.880

EFFECT OF NON-IDEALITY

The assumption $K_a \approx K$, made here, is equivalent to taking the liquid mixture to be ideal, so that $a = X$ for each component. No activity data are available for this mixture, but clearly it is *not* ideal; since methanol forms azeotropes both with C_4s and with MTBE while these latter do not form azeotropes with each other (see Table 2.3 on page 19), the activity of methanol, in particular, must be greater than its mol fraction in the liquid mixture, at least near the mol ratios for azeotrope formation. This factor in isolation would imply $K>K_a$; and the effect is little modified by the other components, because their behaviour is more nearly ideal. Similar liquid mixtures containing methanol, such as methanol/hexane, show methanol activities more than ten times greater than the mol fraction. This implies that K could be similarly greater than K_a; and that, without measured activity data, we can obtain only an extreme lower limit for K, rather than a representative value, for these design calculations.

Fortunately, however, what ultimately concerns us here is not K but X. At the high conversions of present interest, X is very insensitive to the value of K. Accordingly, however much larger that K_a K may be, the X-curve of Figure 3.2 will, in the region of interest, be displaced no more than 2 or 3% upwards. In the context of the reactor design problem posed in Chapter 3, the existing curve of Figure 3.2 thus leads to a conservative, but not unrealistic, estimate of the maximum allowable reactor temperature at which a given (high) conversion can be achieved.

REQUIRED CONVERSION

The essential requirement of the reactor, as stated in Chapter 2, is to provide a product in which the MeOH/C_4 mol ratio does not exceed about 0.095. With a feed precisely equimolar in M, B and inert C_4s, the relative numbers of mols of each of these species in the reactor outflow at conversion X is as follows:

M	$1 - X$
B	$1 - X$
Inert C_4s	1

Hence mol ratio M / total C_4s $= (1 - X)/(2 - X)$

If this is not to exceed 0.095,

$$1 - X = 0.190 - 0.095\,X$$

$$X \geq \frac{0.81}{0.905} \geq 0.89$$

In practice, however, M would be kept in small excess in the feed. The restriction placed on the M/C_4 ratio in the reactor outflow limits this excess of M in the feed to about 9.5%. To accommodate flow control fluctuations it would be wise to set the excess of M at 5% nominal. Relative numbers of mols in the outflow for this mean value of excess methanol in the feed are then:

M	$1.05 - X$
B	$1 - X$
Inert C_4s	1

Hence ratio M / total C_4s $= (1.05 - X)/(2 - X) \leq 0.095$

$X \geq 95\%$

Figure 3.2 then indicates 70°C as the maximum reactor outflow temperature.

While these conclusions are sufficiently assured for use in a feasibility study of the proposed process, it must be pointed out that (as is commonly found by designers), much of the physical data on which they depend is uncertain (or indeed lacking). Thus, besides the lack of activity data for the liquid components, widely different values for the specific heat, $C^\circ_{p\,(l)}$, of isobutene, for example, are quoted in different literature sources. Experimental measurements and verifications would certainly be required before this design scheme was further proceeded with.

ESTIMATION OF ADIABATIC TEMPERATURE RISE

Suppose that, with the feed ratios as reported above, the feed stream is supplied to a thermally insulated reactor at 25°C. The reaction is assumed to take place at 25°C with the enthalpy change $\Delta H^\circ_{R,298}$ and the products then to be heated up to their ultimate temperature by means of the reaction heat evolved. Hess's Law tells us that the temperature calculated in this way will be correct, although obviously the reaction does not actually take place at 25°C.

The quantities of material in the product stream are (per mol of isobutene and methanol fed) as follows:

Material	Mols	Heat capacity, (J K^{-1})
Isobutane	1	133
Isobutene	$1 - X$	$162\,(1 - X)$
Methanol	$1 - X$	$162\,(1 - X)$
MTBE	X	$190\,X$
Total		$391.5 - 68.5\,X$

So a heat balance on the above process gives, since X mols of MTBE have been produced:

$$-(-45{,}460X) = (391.5 - 68.5X)\,(T - 298)$$

$$T = 298 + 45{,}460X\,/\,(391.5 - 68.5X)$$

Values of the temperature reached by a feed entering at 298°K, due to the adiabatic temperature rise, are given in the table below for a number of values of X and are also plotted on Figure 3.2.

X	0.0	0.2	0.4	0.6	0.8	1.0
T, K	298	322.1	349.9	375.8	406.0	438.7

The curve representing these data on Figure 3.2 meets the equilibrium conversion curve at about 80% conversion and 400°K, beyond which no reaction is possible in this adiabatic system. This limiting conversion is too low for Scheme A of Chapter 2 for convenient product separation. On this ground alone, therefore, some means of removing reaction heat is essential.

APPENDIX II — SOLUTIONS TO EXERCISES

EXERCISE 1.2

You are asked to design a plant to produce cyclohexanol/cyclohexanone mixture by oxidising cyclohexane with air. Prepare a Hazard Sheet for cyclohexane.

SOLUTION: PRELIMINARY HAZARD STATEMENT:CYCLOHEXANE

Properties

Formula:	C_6H_{12} MW:84
A mobile liquid, bp 80.7°C, mp 6.6°C	
Vapour pressure:	60 mm/14.7°C; 100 mm/25.5°C
SG:	0.78
Miscibility:	Insoluble in water
Flash point:	–20°C
Explosive limits:	Lower, 1.3% (corresponding to a VP of 10 mm which is exerted at –15.9°C) Upper, 8.4% (=VP of 64 mm/16°C)
Autoignition temperature:	245°C
Toxicity:	Toxic when inhaled or digested. Degreases and permeates skin. LD50; oral, mouse at 1297 mg kg^{-1} corresponding to c. 100 g per adult human.
Exposure limit:	300 ppm v/v in air.

IMPLICATIONS FOR DESIGN

Spillages, even as cold liquid, are volatile, forming vapour clouds which are heavier than air. These may readily be within explosive limits and above flash point. Liquids in process may be close to autoignition temperature in air. Fire hazard is thus severe and there is risk of unconfined vapour cloud explosion.

Liquid cyclohexane is lighter than, and immiscible with, water so that spillages onto wet floors or into drains may spread fire hazards and ignite some distance from the original point of spillage.

Fire fighting measures: foam, dry powder, carbon dioxide or vaporising liquids.

DESIGN PRECAUTIONS

High standards of equipment and piping design and maintenance are necessary to stop leaks, especially from flanges subject to strain on heating and cooling, and at pump and valve glands and seals.

Ingress of air to storage tanks, etc should be prevented by inert gas blanketing. Facilities for safe relief and blowdown of vessel contents in an emergency are required. Vents should be flame-trapped.

No ignition sources are to be allowed within a secure perimeter fence. Electrical installations to accord with BS 5345 Part 1 1976 (Zone 2). Layout to have adequate separation from storages and other plants. Locate downwind of other plants. Provide peripheral and in-plant fire fighting access.

Plant within the fence to be a restricted area with controlled admission. No matches, lighters, static generating footwear or sparking operations allowed. Plant maintenance and modification to be subject to permit and prior approval by designated management. All personnel require special training.

EXERCISE 2.1

MTBE Scheme B allows recirculation of methanol/MTBE azeotrope to the reaction stage. More methanol in the reactor outflow means less MTBE isolated as product. How might this azeotrope be broken? Devise an alternative separation procedure and draw a preliminary block flow diagram for it.

SOLUTION

An alternative separation procedure could be based on water washing.

EXERCISE 2.2

The C_4 stream in Case Study 2.1 originates from the primary distillation column of a steam naphtha cracker. What impurities other than C_4s are likely to be present? Where will these impurities emerge from the proposed separation process? Will they affect performance significantly?

SOLUTION

As remarked under 'Possible impurities (c) From impure reactants', 'Other possible impurities in the isobutene stream are mainly C_4s together with traces of C_3s and C_5s'.

Note that C_3 and C_5 homologues of all the C_4 constituents will probably appear and behave as do their C_4 counterparts; ie MTBE will contain traces of methylpropyl ether and methylpentyl ether unless these are separated in the tops and bottoms of product distillation.

EXERCISE 3.1

Construct a preliminary block flow diagram for benzene hydrogenation, tabulating important features of the reaction and product separation stages and writing a process outline.

SOLUTION

Figure A3.3 is a preliminary block flow diagram. The accompanying process description reads:

'Benzene is received into storage in block 100 from where it is pumped for preparation for reaction in block 200. Benzene and hydrogen are passed to block 300 where the reversible exothermic reaction is carried out, temperature being limited to 175°C so that the residual benzene content is 100 ppm at most. Excess hydrogen used to assist conversion is separated from cyclohexane product in block 400. The product is cooled, depressurised and passed to storage in block 600.'

EXERCISE 4.1

Table 4.1 specifies three condensers to reduce the cyclohexane content of excess hydrogen, items E302, 303 and 304, followed by a product cooler E305. Would a single item be satisfactory? What further attention should be paid to the function and design of these items as the flowsheet is developed?

SOLUTION

Four items are specified rather than one to improve energy efficiency and product yield. Hydrogen saturated with cyclohexane vapour enters E302 at reactor temperature, say 150 – 200°C. In the preliminary flowsheet the latent heat of condensing cyclohexane is rejected to cooling water. This is wasteful, since E302 could be designed as a waste heat boiler raising steam at the highest practicable pressure for general plant use. Hydrogen exit temperature would then be such that the vapour pressure of cyclohexane is still high. This temperature should be reduced before recirculation to reduce energy consumption by circulator C201 and to prevent condensation in it. E303 is therefore a cooler/condenser which can conveniently act as a feed water heater for E302.

A portion of the recirculating hydrogen and inerts is vented as stream 11 to a fuel gas main, inevitably saturated with cyclohexane, causing loss of valuable product. E304 reduces this by further cooling below that achievable in E303.

Product leaves the reactor in stream 4 at reactor temperature for product storage T601 where, for safety reasons, it should be stored at as low a tempera-

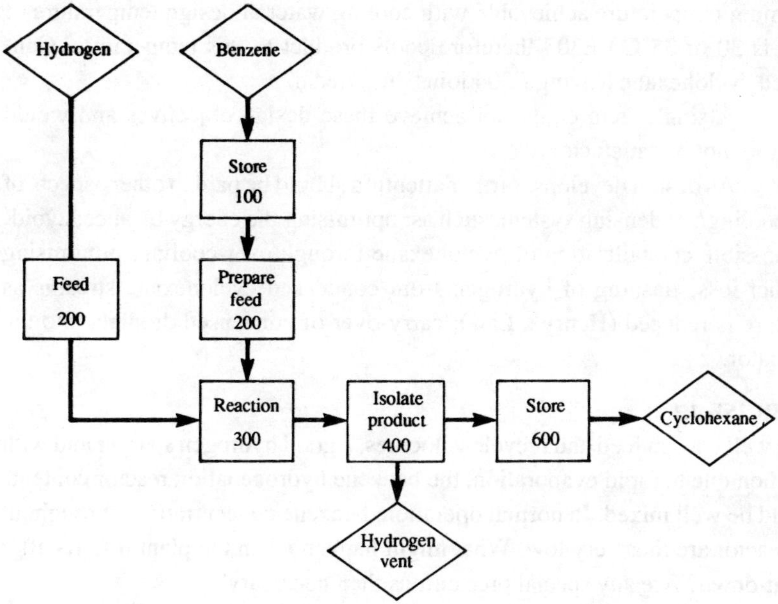

	Reaction block 300	Isolation block 400
Objective	Hydrogenate benzene to 99.99% conversion.	Isolate hydrogen-free cyclohexane.
Constraints	Equilibrium limits maximum temperature to 200°C.	Henry's Law controls hydrogen solubility. Cyclohexane VP governs loss in H_2 vent.
Tentative operating limits	Reaction temperature 175°C. Excess hydrogen minimum at 0.5%. Minimum total pressure 10 bar.	Atmospheric pressure and 35°C maximum at product outlet.
Special problems	Heat removal. Kinetics.	Hydrogen released every time pressure changes.

Figure A3.3 Cyclohexane production: block flow diagram as a working paper.

ture as practicable. (Crystallising point of pure cyclohexane is 6.6°C and minimum temperature achievable with cooling water at design temperature of 25°C is 30 or 35°C.) E305 therefore cools product to this temperature. Condensed cyclohexane leaving E304 joins this stream.

A single item could not achieve these design objectives and would therefore not be satisfactory.

As design develops, further attention should be paid to other aspects of this cooling/condensing system, such as: optimising the energy balance; avoiding possible crystallisation of cyclohexane through over-cooling; minimising product loss; flashing of hydrogen from condensed cyclohexane streams as pressure is reduced (Henry's Law); carry-over of condensed droplets into the circulator.

EXERCISE 4.2

With well chosen feed and recycle velocities, a good hydrogen sparger and with agitation due to rapid evaporation, the benzene hydrogenation reactor contents should be well mixed. In normal operation, benzene concentrations throughout the reactor are thus very low. What might happen when the plant restarts after a shut-down? Are any special precautions then necessary?

SOLUTION

If the reactor should be refilled from the normal benzene supply tank, it will contain far more unreacted feed than normally. The initial rate of reaction at low temperature will be very low, building up as temperature increases in a self accelerating manner; ie there will be a runaway reaction at a rate greater than design levels, leading to excess evaporation and vapour flow with liquid and catalyst carry-over into the condensing system and product tank. This can be avoided if the reactor is refilled with cyclohexane, preheated to reaction temperature and hydrogen and benzene feed restarted slowly while monitoring the condenser loads as a measure of reaction rate.

EXERCISES 6

SOLUTIONS

FLOWSHEET 6.1

(1) Securing high overall conversion (product quality) requires low reactor exit temperature; elsewhere in the reactor, higher temperatures are desired for faster reaction and economy of reactor volume. These aims can be achieved by

dividing the catalyst into several successive beds with inter-stage cooling, operating at successively lower outlet temperatures. More generally, the arrangement shown avoids the presence of cooling tubes within the catalyst mass and the region of three-phase flow (see Answer (2(b)), Flowsheet 6.2, below).
(2) Both outlets from the reaction system are flow controlled to maintain the reactor pressure; the vapour outlet by PC1 on the hydrogen vent stream and the liquid outlet by LC1, which keeps the top catalyst bed wetted.
(3) Since the hydrogenation reaction removes hydrogen from the recirculating stream, the latter is richer in inerts than the feed. This reduction in hydrogen partial pressure reduces the reaction driving force. The effect on conversion must be allowed for. At constant temperature and total circulation rate, however, heat removal capacity is not affected.
(4) At steady state, 'what goes in must come out' so that, at constant operating conditions, hydrogen vent rate equals hydrogen feed rate minus hydrogen reacted. Inerts out also equal inerts fed. Purge rate must be increased for increased hydrogen feed rate. This reduces the inert content of the recirculating stream but increases rate of hydrogen loss to fuel gas.
(5) Flowsheet 1 indicates that E1303 and E1313 are at a higher level than the reactor. Cyclohexane condensate can then return to the reactor by gravity, assisted by the density difference between the hot hydrogen/benzene/cyclohexane mixture in the reactor and colder condensed cyclohexane in the downcomers.
(6) Sparging hydrogen into the reactor produces vigorous agitation, producing sufficient back-mixing to raise the mixture of cold feed and hot reactants to catalyst initiation temperature.
(7) Product flows direct from V1401 to the single storage T1601. Good practice avoids simultaneous make-and-take at the same tank. While on-line analysis is valuable for process control, sales specifications usually require additional tests. The flowsheet should therefore specify up to four test tanks holding a shift's or day's make before product is put to final storage. One tank is under test, one being filled and the third being pumped out. A fourth tank contains any off-specification material for subsequent blending or reworking. At the line diagram stage, the line should enter T1601 above liquid level with its open end then submerged to prevent splashing and build-up of static electricity.

Cyclohexane freezes at 6.5°C so that, in the UK, heating is required in winter.

FLOWSHEET 6.2

(1a)Liquid level in reactor R2301 is determined by the difference between inflow and outflow. Inflows are:

(i) benzene feed from P2102, plus

(ii) condensed cyclohexane returned by P2316 from E2313 and E2314. The outflow is cyclohexane vapour in the excess hydrogen stream to E2313 and E2314. This is fixed by hydrogen flow, reactor temperature and total reactor pressure. FC1 controls benzene feed to the reactor. The only freedom remaining is to regulate condensed cyclohexane return (ii) by LC1, maintaining desired reactor operating level. The remaining condensate flows to V2401. Careful consideration would have to be given to the question whether this gives sufficient control to meet all probable variations in operating conditions.

(b) Continuous operation of Process 2 may be interrupted should P2316 fail occasionally. An installed spare should be specified here and wherever else failure could lead to an unacceptable process interruption. If E2313, E2314 and V2315 were at a sufficient elevation, as in Process 1, condensate could be returned to the reactor by gravity. This is likely to be a more reliable arrangement.

(2a)Heat of reaction is removed from R2301 in three principal ways:

(i) by the internal coils of the waste heat boiler supplied by pump P2309;

(ii) by evaporating cyclohexane into the recirculating excess hydrogen stream; and

(iii)by the small cooling effect of returning condensate from E2313 and E2314.

Most heat is thus rejected into the waste heat boiler raising steam for profitable use.

(b) In Process 2, waste heat boiler tubes within the reactor maintain near isothermal operation. This is not the best arrangement for securing high conversion at low reactor volume (see Answer (1), Flowsheet 6.1 above). Moreover, contact of catalyst, liquid and gas may be impeded by the tubes in the reactor. This does not occur in Process 1.

Continued operation of the waste heat boiler in Process 2 depends on the reliability of boiler water circulating pump P2309. No such pump is shown in Process 1, but will probably be needed there also.

(3) Pressure is measured at the coolest available point down-stream of the reactor and maintained by PC1 which restricts the rate of hydrogen purge to the fuel gas main. The whole of the hydrogen recirculating loop is then at high pressure, including V2315. Liquid transfer to the low pressure item V2401 is controlled by LC3.

(4) Cyclohexane in V2315 is saturated with hydrogen at the prevailing low temperature and high pressure. Henry's Law indicates that after passing through the control valve of LC3, to lower pressure, hydrogen solubility is reduced, hydrogen being released and separated in V2401.

(5) Under operating conditions, the reactor contains a low density mixture of hydrogen and liquid whose quiescent volume is below normal operating level. If filled with liquid before start-up, the excess is entrained into E2313, E2314 and V2315 and hence forward into T2601. If the reactor has been filled with benzene, reaction could be violent. Product is thus likely to be contaminated with entrained catalyst as well as benzene.

(6a)Before start-up, the system should be filled with specification quality cyclohexane and not benzene. A pump and line is required to transfer this from storage and liquid level measurement is required half way up the reactor.

(b) Reaction cannot start unless reactants are at the catalyst initiation temperature, which is probably about 120°C. Flowsheet 6.1 shows a preheating coil within the reactor, but there appears to be no such item on Flowsheet 6.2. A start-up heater should be provided in the liquid feed line to the reactor or in the reactor itself.

(7) Steam outflow from drum V2308 should be controlled at constant pressure and hence at constant temperature. Changing the set-point varies heat transfer in the reactor and consequently changes its temperature, but sudden reduction will reduce the boiling temperature in the drum, resulting in surging with possible water carry-over into the steam outflow. This could be avoided if the steam outflow pressure (temperature) from V2308 were treated as a control variable; but then the export steam would be of variable quality and hence of little sale value.

Although feed to the drum is of boiler water quality, blow-down facilities are necessary to prevent the build-up of impurities with consequent corrosion.

(8) Inerts are vented from the recirculating gas stream through a branch between V2315 and C2201, restricted by the reactor pressure controller PC1. Recirculating gases are cooled by the water-cooled condenser E2314 so that their design temperature is at best, say, 30°C, unless the available water is unusually cold. At this temperature cyclohexane vapour pressure is 120 mmHg, leading to cyclohexane loss to the vent which could be recovered, if economic, by refrigerated cooling or absorption.

If cyclohexane condenses in the feed line and is entrained into the hydrogen circulator the mechanical performance of this item may be affected. A mist removal pad should be installed.

FLOWSHEET 6.3

(1) The first reactor has no liquid overflow, nor is any condensate returned nor cyclohexane recycled. Liquid level is maintained by balancing cyclohexane evaporation rate against benzene feed. Evaporation rate depends on:

(a) reactor temperature, which determines cyclohexane vapour pressure and hence its partial pressure in the gas stream; and

(b) reactor pressure, fixing the partial pressure ratio between cyclohexane and excess hydrogen. Reactor temperature is affected not only by evaporative cooling but by heat removed in the waste heat boiler heat exchanger E3303. Maintaining correct liquid level thus apparently requires correct manipulation of:

(i) steam pressure in V3308, the drum of the waste heat boiler;

(ii) reactor pressure; and

(iii) gas recirculation rate.

Quite a task! Control of Process 1 looks much simpler.

(2) The heat of reaction is removed by:

(a) E3303 into the waste heat boiler system, raising steam; and

(b) evaporating cyclohexane into recirculating excess hydrogen. This is then cooled in the boiler feed water preheater E3313 and water-cooled exchanger E3314.

(3) A disadvantage of this arrangement appears to be the difficulty in maintaining liquid level and at the same time varying reactor temperature independently, according to catalyst performance.

(4) In this process the catalyst is suspended in the circulating liquid in the reactor; as a consequence the heat exchanger E3303 also operates with entrained catalyst, presumably on the tube side. After prolonged use Raney nickel disintegrates and forms sludge. This is likely to lead to fouling of heat transfer surfaces by catalyst fines and will jeopardise continuous pump operation, especially since the off-take is from the base of the reactor. If the direction of flow were reversed, fewer catalyst fines might be entrained. Catalyst suspension should pass through the tubes so that fouling will be minimised, or at least fairly readily removed in periodic cleaning.

(5) R3312 is a vapour phase reactor containing a robust supported nickel catalyst through which all unchanged reactant from the first reactor passes. If the Raney nickel in the first reactor is partially poisoned, the vapour outflow will contain unreacted benzene which will react in R3312. This is therefore a safeguard of product quality. There is no independent control of its operating conditions, although conversion may be enhanced by raising the hydrogen

partial pressure in the recirculating gas, ie by increasing hydrogen feed rate. Its temperature rise will indicate the amount of conversion it performs.

(6) V3401 is a phase separator on the product outflow from separator V3304. Cyclohexane in V3304 is saturated with hydrogen at the prevailing high pressure. After passing through the control valve of LC3, Henry's Law indicates that hydrogen solubility is reduced at the lower pressure of V3401, where hydrogen is released and separated. Its hydrogen outlet should be controlled to a pressure higher than the fuel gas main. Liquid outlet to storage should be level controlled. Further slight hydrogen release will occur in this outlet.

(7) C3201 is a hydrogen recirculator accepting vapour from separator V3304 in equilibrium with condensed cyclohexane leaving the water-cooled condenser E3314. With a design temperature for cooling water of, say, 25°C, liquid temperature is likely to be, say, 30°C, at which cyclohexane exerts a vapour pressure of 120 mm. The vapour therefore contains significant amounts of cyclohexane which condenses as its temperature falls. It is essential to avoid liquid entrainment into the circulator inlet, to prevent damage to the working parts of the machine. A spray separator/mist eliminator should be installed before the inlet.

(8) See Answer (3), Flowsheet 6.2.

(9) A reciprocating compressor should never be throttled on its outlet; so the arrangement of FC2 is dangerous for this type of machine. Control should use a by-pass from outlet to inlet.

EXERCISE 7.1

Consider the above operations of ion exchange and regeneration and prepare a flowsheet and plant item list to perform them. Item sizes are not required.

SOLUTION

The flowsheet Figure A7.3 on pages 116 – 117 and plant items listed in the key achieve the design objectives represented by the solutions proposed on page 68 to problems (1) to (7) in the operation of regeneration.

EXERCISE 9.1

Draw an engineering line diagram to show the process pipelines and ancillaries required for the feed section of an alkene dimer plant starting at the inlet to the transfer pump P9101 and ending where feed leaves the interchanger H9203. No line diagram additions are required for the feed preheater item H9204 or the reactor R9205.

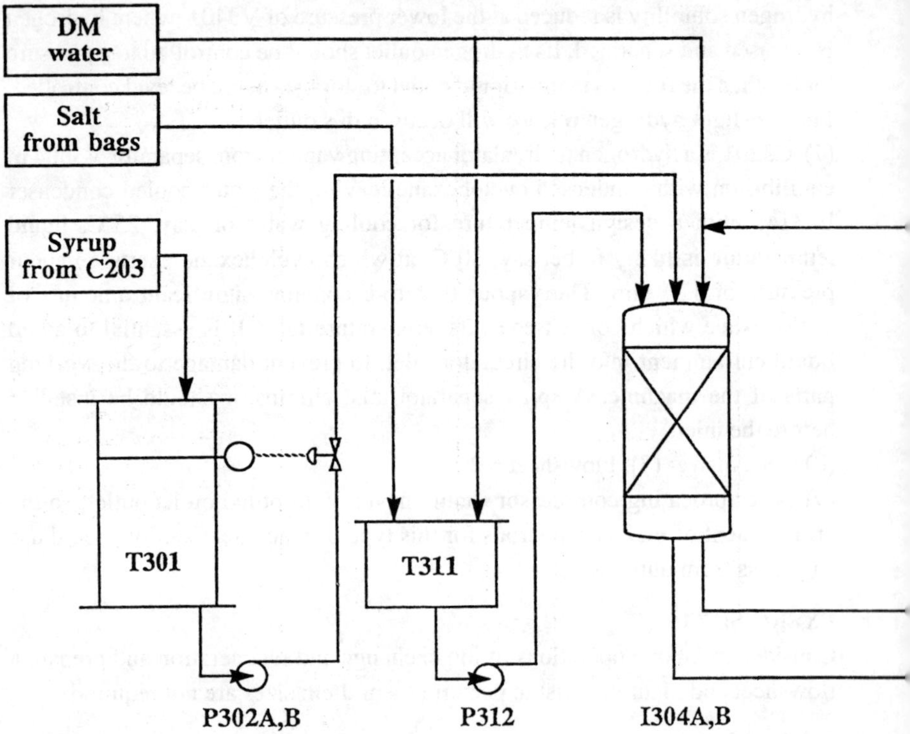

Figure A7.3 Corn syrup ion exchange flowsheet.

SOLUTION

Important features relating to Figure 9.1, pages 80 – 81, include:

(1) Control of retention time in the reactor by flow control of feed at a cool position on the outlet of P9101.

(2) The reactor contents are maintained in the liquid phase by pressure control downstream of the reactor at a cool point after H9203.

(3) Pumps P9101 and P9202 have installed spares. P9202 pumps have restric-

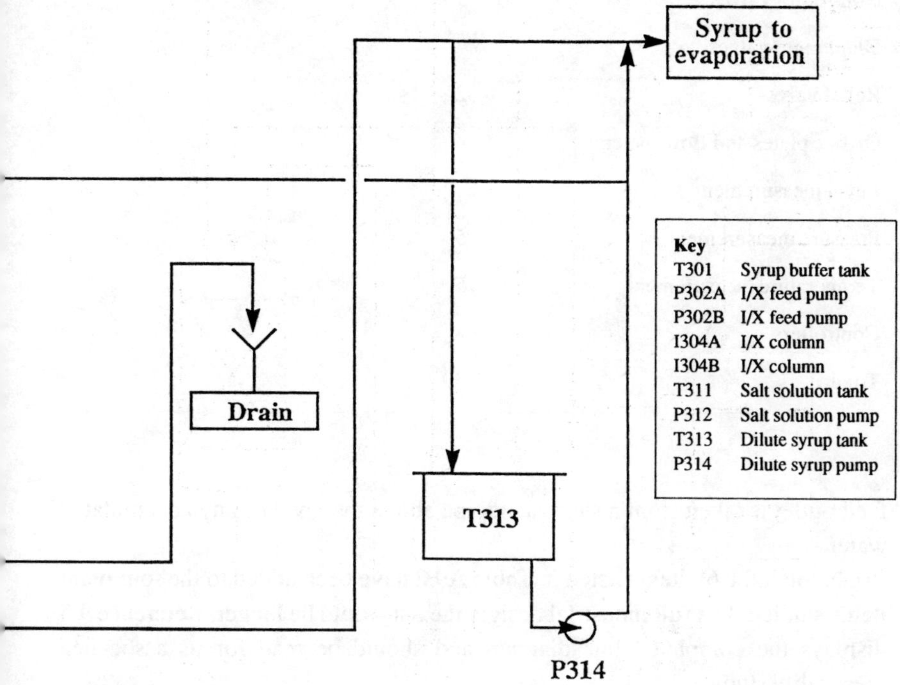

tion orifice plates to guard against excessive pressure during possible dead heading (Reference 9.6).

(4) Pressure relief devices are provided on the feed transfer line from bulk storage, on T9201, and on the shell side of H9203.

(5) There is controlled nitrogen addition and venting on T9201 to prevent explosive hydrocarbon/air mixtures forming. The hydrocarbon feed inlet is extended below the liquid surface to prevent static electricity generation. The

Table A9.2
Alkene dimer line diagram: valves and instruments summary

Item	Number
Manual valves and cocks	40
Non-return valves	5
Diaphragm valves	4
Relief valves	2
Orifice plates and flow meters	4
Level measurement	1
Pressure measurement	5
Temperature measurement	5
Controllers	3
Total	69

feed outlet is taken from a short standpipe above the level of any accumulated water.

(6) In total, the 69 items listed in Table A9.2 have been added to the four plant items studied. In a full industrial design, the list would be longer. Reference 9.5 displays the completed line diagram and should be read for its associated operability study.

APPENDIX III — CHEMICAL SYNONYMS, ABBREVIATIONS AND SYMBOLS USED IN THE TEXT

CHEMICAL SYNONYMS AND ABBREVIATIONS

Alumina	Aluminium oxide
B	Isobutene (in Appendix I)
BS	British Standard
CS	Carbon steel
DMT	Dimethylterephthalate
EDC	Ethylene dichloride, 1,2 dichloroethane
ELD	Engineering line diagram
FC	Flow controller
I/X	Ion exchange
LC	Level controller
M	Methanol (in Appendix I)
MTBE	Methyltertiarybutyl ether
Naphtha	A petrol-(gasoline-) like mixture of liquid hydrocarbons of boiling point range around 50°–200°C, obtained in the fractional distillation of crude petroleum
Oligomers	Polymerised substances comprising a few molecular units
PC	Pressure controller
Raney nickel	A very finely divided form of metallic nickel used as a hydrogenation catalyst
SS	Stainless steel
TA	Terephthalic acid
TEL	Tetraethyl lead
VLE	Vapour-liquid equilibrium

MATHEMATICAL SYMBOLS

a	Activity of substance specified by subscript
C°_p	Standard molar heat capacity of species indicated by further subscript
ΔC°_p	Standard molar heat capacity change for chemical reaction

ΔG_R	Free energy change in chemical reaction
ΔG°_R	Standard molar free energy change in chemical reaction
ΔG°_f	Standard molar free energy of formation of species indicated by further subscript
ΔH°_R	Standard molar heat (enthalpy change) of reaction
ΔH°_f	Standard molar heat of formation of species indicated by further subscript
ΔH_R	Heat of reaction
ΔH_V	Molar heat (enthalpy change) of vaporisation of species indicated
K	Equilibrium constant for chemical reaction based on mol fractions of participating species
K_a	Equilibrium constant for chemical reaction based on activities of participating species
ln	Logarithm to base e
n	Exponent in cost versus plant item capacity relation (Chapter 10)
N	Number of theoretical plates in a distillation column
P	Total system pressure
P_{crit}	Critical pressure
P_{max}	Maximum acceptable operating pressure
p_1	Partial pressure of component 1 in mixture
p^*_1	Saturation vapour pressure of species 1
R	Universal (molar) gas constant
S	Number of theoretical distillation stages required to effect a given degree of separation of components of mixture
T	Absolute temperature
T_{crit}	Critical temperature
T_{max}	Maximum acceptable operating temperature
T_{op}	Operating temperature
X	Fractional conversion in chemical reaction
x_1	Mol fraction of component 1 in liquid phase
α	Relative volatility of components of mixture

INDEX